Human Factors and Design

Yuri Spirochkin

Human Factors and Design

Springer

Yuri Spirochkin
International Consultancy and Analysis
Agency "Aviation Safety"
Moscow, Russia

ISBN 978-981-19-8831-8 ISBN 978-981-19-8832-5 (eBook)
https://doi.org/10.1007/978-981-19-8832-5

This Springer imprint is published by the registered company Springer Nature Singapore Pte Ltd.
The registered company address is: 152 Beach Road, #21-01/04 Gateway East, Singapore 189721, Singapore

Preface

Two-thirds of flight incidents, accidents, and catastrophes, as statistics show, are due to "human factors". The cause of the disaster at the *Fukushima Daiichi* nuclear power plant in 2011 was a huge tsunami wave, and humans, as it seems, were not guilty in it. My point of view is that human factors are present in emergency events of any origin, and only the assessment of its role in the situation differs depending on what this term means. I am of the opinion that human is 100% responsible for all accidents and catastrophes that occur with technical systems. There are two methods to substantiate it: simple and complex.

The first method is in one sentence: Technical systems are human-made, so any adverse event happening to them is due to a fault of human, because he (or she) has assumed the role of the creator of the "second nature". If such simple substantiation does not suit the reader, there is the second, complex method. To get acquainted it, you will have to read this book. It covers various manifestations of human factors when interacting with potentially dangerous technical systems: nuclear power plants, air- and spacecraft, ground vehicles, etc. This interaction begins in the design stage.

The idea for the book arose out of the desire to find a common ground between the industries that are important to safety and are closest to me: aerospace engineering and nuclear power.

My first specialization obtained at the N. E. Bauman Moscow State Technical University is the design of space rockets and spaceships. From 1975 to 1994, I worked in the Scientific and Production Association (in Russian, it sounds as "Nauchno-Proizvodstvennoe Ob'edinenie", or NPO) *Energia.*[1] The main project in which I was lucky enough to take part was design of the space transport system *Energia-Buran.*[2] Within the framework of this project, I collaborated with NPO *Molniya* and

[1] Now, Joint stock company (JSC) S. P. Korolyov Rocket and Space Corporation *Energia.*

[2] A Soviet analog of the American *Space Shuttle* system.

TsAGI.[3] In the early 1990s, the project was closed like many other state programs, but opportunities for independent engineering activities were opened up. We, a handful of employees of NPO *Energia*, organized under the auspices of the Association "Aviation Safety Foundation of the USSR" a research and engineering company *Dynamika*, and I was its director for several years. We were developing computer programs for ensuring safety of air and space vehicles; also, we were beginning to carry out numerical simulations of accidents of cars and railway trains. Thanks to my colleagues from TsAGI and the membership in the Association, I got acquainted activities of some Russian aircraft design bureaus. In the early 2000s, the situation with the private engineering business in Russia worsened. In 2005, I signed an employment contract with the state institution NIKIET,[4] and the next 15 years of my career were connected with Russian nuclear power industry. The results of this work are described in the book (Spirochkin 2019). While working in nuclear power, I did not break ties with aerospace industry. For several years until March 2022, I held a honorary position of Chairman of the Board of Directors at JSC Consortium *Space Regatta*. From 2020 to the present, I carry out duties of an expert at the International Consultancy and Analysis Agency *Aviation Safety*, Moscow.

What is the common ground between aerospace technology combining aviation and cosmonautics and nuclear power industry? Currently, engineers design spaceships with nuclear power supply and propulsion systems on nuclear thrust. They develop projects of Lunar or Martian bases functioning due to energy from atom reactors, etc. However, the most common characteristic is that for these industries equally high safety requirements are inherent, and these requirements cover the influence of human factors. It is very interesting to analyze such influence in the context of safety.

The matter is rather complex: human controls a technical system, his (or her) involvement in its operation determines in some cases the need for human life support, but, on the other hand, human errors, negligence or evil intentions, can turn the system into a threat with disastrous consequences. Still, the first aspect of the influence in question is that human creates technical system. During this process, in the design stage, human must ensure the safe operation of the system, taking into account all components of possible risk, including those caused by human factors itself. The book is about how designers do it now and what should they do for greater safety in the future.

The first edition of this book has been published in Russian in the Saint Petersburg publishing house *Right Print* at the beginning of autumn 2020. Almost simultaneously, an authentic electronic version was released in the Moscow publishing house *De'Libri*. This English-language edition is actually the second edition of the book,

[3] NPO *Molniya* is a company that created the Soviet shuttle *Buran*. TsAGI is the Russian abbreviation of "Tsentralnyi Aero-Gidrodinamicheskii Institut". Current full name of this organization is the State Scientific Center—Federal State Unitary Enterprise "Prof. N. E. Zhukovsky Central Aerohydrodynamic Institute".

[4] NIKIET is the Russian abbreviation of "Nauchno-Issledovatelskii i Konstruktorskii Institut Energo Tekhniki". Now, the name of this organization is JSC N. A. Dollezhal Research and Design Institute of Power Engineering.

in which the noted inaccuracies of the original issue are corrected, the text of a number of paragraphs is revised and supplemented, and some images are updated. The bibliography is significantly expanded by adding new sources.

Saint Petersburg, Russia
June 2022

Yuri Spirochkin

Summary

Human involvement in the life cycle of complex technical systems: nuclear power plants, air- and spacecraft, ground vehicles, etc., creates a specific phenomenon known as “human factors”. Existing methods for analyzing this phenomenon and managing it for safety purposes relate mainly to operation of systems. They consider certain features of human nature and allow an assessment of human behavior in interaction with machines, including human errors. The author extends analysis to the design stage of a system. The analysis reveals gaps in the existing approaches to human factors in this stage and identifies possibilities to increase safety from the standpoint of designer. These possibilities cover a shift in the design paradigm, and improvements in the methods used for solving engineering problems and managing design activities. Such a shift and improvements are necessary because of the tightening of safety requirements, which is being carried out largely due to accidents and catastrophes that have caused a great resonance in society.

The book is intended primarily for designers, analysis engineers and researchers, as well as managers at organizations working in aviation and space engineering, nuclear power and transportation, or creating other potentially dangerous machines and infrastructure facilities. It can be useful also for students of technical universities and graduate students. Factual data on achievements and ideas in hi-tech industries can attract the attention of other technology-fascinated readers.

Contents

Chapter 1
Introduction

1.1 History and Definitions of Human Factors

It is not easy to ascertain when the term *human factors* appeared as well as to find a precise definition of it. The author is aware of the book Chapanis et al. (1949), published in 1949, in the title of which this term has been used apparently for the first time. However, what it means was not clearly explained in the book. The area covered by this term was outlined as "the behavior of human beings while working with machines or instruments". In the following years, researchers and representatives of industry were making efforts to transform "human factors", originated from hi-tech weaponry, mainly military aviation, into a general engineering discipline applicable in any technology and were using it to create human–machine systems—see for example Chapanis (1965), Wirstad (1979). Human–machine systems are defined as those in which humans have a control or monitoring function. The discipline "human factors" was at the junction of psychology, physiology, physics, design and engineering analysis, and it was being developed because designers needed precise knowledge of human capabilities and limitations in order that machine parameters as well as operating requirements meet them.

In the course of the development of this discipline, it acquired synonymous names: *human factor engineering*, *human engineering* and *ergonomics* (Swain and Guttmann 1983). The goals of applications of this discipline are to reduce human errors, increase productivity and enhance safety and comfort with a specific focus on the interaction between human and machine (Wickens et al. 1997). According to the *Big Soviet Encyclopedia*, the main and most extensive reference publication in the Soviet Union (until 1991), a right combination of human abilities and machine capabilities significantly increases the efficiency of human–machine system and contributes to its optimal use.

The interchangeability of the terms *ergonomics* and *human factors* is reflected in the international standard (ISO 6385:2016):

Y. Spirochkin, *Human Factors and Design*,
https://doi.org/10.1007/978-981-19-8832-5_1

> **2.3 ergonomics | human factors** scientific discipline concerned with the understanding of interactions among human and other elements of a system, and the profession that applies theory, principles, data and methods to design in order to optimize human well-being … and overall system performance.

Unfortunately, the existence of several synonyms, even legitimized by the international standard, does not clarify the phenomenon of human factors. Considering it only as a scientific discipline reduces its supposed content. This is not only an opinion of the author. The problems of terms and definitions in the area in question are discussed, e.g., in the report Licht et al. (1989). This report exposes "the apparent differences among human factors, human factors engineering, and ergonomics". According to one of the sources cited in it: Air Force Systems Command Design Handbook 1–3. Human Factors Engineering (1977), "human engineering is not synonymous with human factors. The term "human factors" is more comprehensive, covering all biomedical and psychosocial considerations applying to man in the system. It includes not only human engineering, but also life support, personnel selection and training, training equipment, job performance aids, and performance measurement and evaluation". Definition of human factors given in the multi-volume handbook (Salvendy 1987) covers "the role of humans in complex systems, the design of equipment and facilities for human use, and the development of environments for comfort and safety". From later publications, we should mention the book chapter (O'Brien and Meister 2002) in which the concept we are considering is characterized indirectly—through "the technical merit and operational effectiveness and suitability of any human–system interface". Human factors testing and evaluation should include "training, manpower and personnel, health hazards, and perhaps others". Recently released handbook of the U.S. Fire Administration (USFA 2020) defines "human factor" (in the singular) as "any physical or cognitive property specific to an individual or any behavior specific to humans that may influence the functioning of mechanical and technological systems." The text of the handbook following this definition contains the term "human factors", i.e., in the plural.

Often, the concept of human factors is applied to describe the possibility of erroneous or illogical decisions of humans in certain situations and as the designation of a common source of accidents and catastrophes caused by such decisions. About 70% of all plane crashes are caused by "pilot errors" ranging from unreliable manual controls to inadequate flight planning (Salvendy 1987). Such errors are mainly unintentional actions of human operator that go beyond acceptable limits. At the same time, manifestations of human factors include decisions that are logically substantiated, but are made on the basis of incorrect or incomplete information, chaotic human behavior in emergency situations, as well as deliberate malicious actions that lead the system to an uncountable state. These deliberate malicious actions cover among others terrorism.

We will use the term *human factors* to characterize the features of human nature and individual or group behavior of humans who are part of human–machine systems or are involved in the life cycle of a technical system by another way. The concept of *technical system* seems to be more general than *human–machine system* since it includes systems in which humans lose the control of the machine in an abnormal situation, or unmanned systems operating in the human habitat. We will consider such technical systems as nuclear power plants (NPPs), air, space and aerospace vehicles, their traffic control centers, ground vehicles, infrastructure objects, etc.

1.2 Negative and Positive Aspects of Human Factors

Due to the continuous improvement of technical systems, the probability of failures of their mechanical or electronic components lowers. The safety targets established for transportation systems under design, energy facilities or civil infrastructure correspond usually to the probability of accident 10^{-6} and that of catastrophe 10^{-7}. These probabilities are ensured through the high reliability of structures, automated control systems and safety systems, as well as the property of inherent safety, characteristic, e.g., for new types of nuclear reactors (Heikkilä 1999). However, this does not improve the overall statistics of accidents and catastrophes—largely due to the influence of human factors embodied in human operator and (or) in other people involved in the life cycle of technical system.

From these statistics over the past 40 years, the following resonant negative events can be mentioned:

- crashes of the supersonic liner *Concorde* near Paris (2000), the Boeing-777 airplane, Flight MH-17, in Donetsk region (2014), the Boeing-737-8 (MAX) airplane, Flight JT-610, near Jakarta (2018) and the same aircraft type, Flight ET302, near Addis Ababa (2019);
- catastrophes of the space shuttle *Challenger* during its launch from Cape Canaveral (1986) and the space shuttle *Columbia* over Texas during the return to Earth (2003);
- the *Three Mile Island* NPP accident (1979), the Chernobyl catastrophe (1986) and the disaster at the *Fukushima Daiichi* NPP (2011);
- collapse of the towers of the World Trade Center in New York (2001).

Despite the differences in immediate initiating events, all these accidents and catastrophes have a common underlying cause: human factors.

In the crash of Boeing-777 close to Donetsk, the accident at *Three Mile Island* NPP, the Chernobyl catastrophe, or the attack of the planes hijacked by terrorists on the World Trade Center, the manifestations of human factors are obvious: humans directly initiated these events. Nevertheless, since the other above events also occurred with the systems made and controlled by humans, their real causes are features of human nature, including inherent human limitations "imprinted" in machines, infrastructure facilities and the processes of their use.

Indeed, the designers of *Concorde* did not calculate that its structure should withstand an impact into the fuel tank of a piece of tire that was torn when hitting a metal object lying on the runway (McEvely 2002), and the employees of the Charles de Gaulle airport did not control over the condition of the runway properly.

The space shuttle *Challenger* was launched in frosty weather. Engineers of the company *Thiokol*, which have created the solid-fuel launch boosters, were aware of the adverse effect of low temperature on the reliability of the rubber seals between their sections. These specialists were insisting to postpone the launch, but managers from National Aeronautic and Space Administration (NASA) of the USA did not listen to them (Presidential Commission 1986). The catastrophe of *Columbia* was caused by the damage of the thermal insulation during the launch and subsequent destruction of the wing under incandescent gas when re-entering the atmosphere (CAIB 2003). In both cases, NASA management, instead of organizing a work to improve the designs of launch boosters or spacecraft, perceived the problems known to specialists as an acceptable risk of equipment failure. Such perceptions of safety were severely criticized later, as well as the NASA's culture of decision-making in the whole. The result was the reform of NASA and the closure of the *Space Shuttle* program.

The investigations of both Boeing-737-8 (MAX) crashes revealed the same causes (KNKT 2019):

- flaws in the newly installed Maneuvering Characteristics Augmentation System (MCAS) designed to help pilots to cope with the stall;
- inadequate certification procedure: safety assessment of MCAS was actually carried out by the aircraft manufacturer itself, not by the regulator, i.e., the Federal Aviation Administration (FAA) of the USA;
- lack of information on MCAS in the pilot manual.

Thus, human factors manifested itself in the errors of MCAS creators, and in the inadequate procedure carried out by the personnel on several levels of the hierarchy, responsible for certification and implementation of this system.

The most hidden is negative role of human factors in the disaster at the *Fukushima Daiichi* NPP. The direct cause of this event was a tsunami wave induced by an earthquake in the ocean off the eastern coast of Japan. An unusually high wave overcame the 6-m protective wall separated the plant territory from the ocean and caused the flooding of the main power supply devices at all four nuclear reactors. The reserve diesel generators were also flooded. The interruption of power supply to the pumps of the cooling system led to the termination of coolant circulation and, as a result, to the destruction of cores in three of four reactors, explosions of the released hydrogen and partial collapse of the reactor buildings. A significant amount of radioactive material was released into the environment. Several meetings of experts at the International Atomic Energy Agency (IAEA) were devoted to the analysis of this accident. In the one of them, the role of human factors was considered,[1] and two errors of the designers of the *Fukushima Daiichi* NPP, which became obvious after the accident, were discussed:

---

[1] IAEA International Experts' Meeting on Human and Organizational Factors in Nuclear Safety in the Light of the Accident at the Fukushima Daiichi Nuclear Power Plant, Vienna, 21–24 May 2013.

(1) the use of statistical data on heights of tsunami waves which did not take into account the probability of very rare, extreme values (the height of the wave on March 11, 2011 was about 15 m and exceeded the statistically substantiated maximum by 3 times);
(2) arrangement of the power supply devices at the lower height marks of the reactor buildings (thus, they were subject to flooding in such events).

The list of accidents and catastrophes caused by human factors in various negative manifestations can be continued.

Of course, human factors have positive aspects. As Stanisław Lem in his philosophical work *Summa Technologiae* pointed out, the systems in which human is included as a control link potentially have greater resistance to adverse external influences and damages, "since the plastic behavior of human often makes it possible to compensate for the violation of functions" (quote from the Russian-language edition of the work (Lem 2019)). Regulatory documents being in use in various industry branches provide for appropriate human actions in abnormal situations. For example, the relevant provision of Russian *Aviation Rules* (AP-25 2009) states: "if a failure condition (functional failure, type of system failure) leads to an emergency …, the Flight Manual should contain recommendations that allow the crew to take all possible measures to prevent the transition of emergency into a catastrophic situation".

If we turn to the regulatory documents of Russian nuclear power industry, we find that the first document that opens their series, (NP-001-15),[2] requires "ensuring the necessary level of qualification of NPP personnel for actions during normal operation and violations of normal operation, including pre-emergency situations and accidents".

As a recent accident in which human factors played a positive role, we can note the emergency landing of the *Ural Airlines* Airbus A321 near the city of Ramenskoye on August 15, 2019. The plane which took off from the *Zhukovsky* airport, heading to Simferopol, collided with a flock of seagulls while climbing. This collision caused both engines to shut down. The pilot landed the plane in a nearby cornfield (see Fig. 1.1) without releasing the landing gear to avoid serious consequences if it could break down. After the landing, the crewmembers extinguished the engines that caught fire and evacuated all 226 passengers; some of them were injured during the landing process (ASN 2019).

However, the development of professional skills of the personnel involved in the operation of technical systems is unlikely to solve all the safety problems associated with human factors. Such a limited approach to accounting for and managing this phenomenon excludes from consideration the errors of creators of the systems, unpredictable behavior of people who use the systems but do not belong to the personnel

[2] These documents (their number currently exceeds 100) have the common name in Russian: "Federalnye Normy i Pravila v oblasti ispolzovaniya atomnoi energii" (FNP) that means *Federal regulations and rules in the field of atomic energy use*. Their designation includes two letters: NP (in Russian "НП"), followed by the document number; the last two digits after the dash indicate the year of issue.

Fig. 1.1 The *Ural Airlines* Airbus A321 after emergency landing on August 15, 2019. Courtesy of Igor M. Konovalov and Anatoly Y. Zhurin

(e.g., passengers) or just are in the area of their work, as well as deliberate malicious actions. The purpose of safety ensuring activities is to minimize the risk associated with harm to the life or health of people, their property and the environment. Taking into account this purpose, the possibility of compensation by the operating personnel for the unreliability of machine components or the adverse events not considered in the design stage is particularly problematic for unmanned vehicles or facilities.

There is a general tendency to tighten the safety requirements to human-made systems in all areas of technology. In nuclear industry, it manifests, for example, in increasing the required seismic resistance of NPP equipment and pipelines, buildings and structures, including the strength of reactor containments. After the accident at the *Fukushima Daiichi* NPP, the regulatory-specified intensity of earthquakes that nuclear facilities shall withstand was increased, the quantities of emergency cooling systems and power supply devices for nuclear reactors were multiplied, etc.

The safety requirements for containments were changed over the past 4 decades even more dramatically. Initially, protective shells were being designed only to retain radioactive materials when the coolant circulation circuit would be depressurized due to internal reasons. Then, in regulatory documents of different countries, the requirements emerged that the reactor containment should withstand the impact of an aircraft with masse of several tons at velocity of 100 m/s, a 20-ton *Phantom*-type fighter at twice the velocity, etc. In modern Russian NPP designs, the fall of a wide-body Boeing 747 aircraft with mass up to 400 t on the reactor building is specified (Birbraer and Roleder 2009).

1.3 The Main Questions Answered by the Book, Its Plan and Intended Readers

Victims and material damage as results of accidents and catastrophes are a high price that society has to pay for the possibility of creating safer technical systems. To reduce this horrid price, it is necessary to consider risk and identify potential for managing risk factors in the earliest stage of the life cycle of any system—during its design. This book considers one category of these factors—human factors. The main questions to be answered are the following:

1. What approaches to human factors in the design stage exist?
2. How fully do they ensure safety?
3. To what extent do these approaches meet modern, more tighten safety requirements?

Above, we defined the term "human factors" and clarified what the concept of "technical system" covers. For an unambiguous understanding of the ideas presented in this book, it remains to explain what the author means by the word "design" which is even included in the title. The established definition of the term (in the form of a verb or noun) can be found, for example, in the Cambridge Dictionary. Along with this, it seems necessary to give a more "technical" explanation.

Design in the sense in which the author understands it means:

(1) a set of documents describing in more or less detail the technical system to be created;
(2) the process of preparing this set and, thus, creating a documented image of the system;
(3) a specific area of human activity associated with the creation of such images, before their embodiment "in metal".

The word "design" is used further in all these meanings, but the latter is predominant. Because of this predominance, the genre of this book can be defined rather as philosophy of design.

The main part of the book that follows this introduction contains four chapters—from the second to the fifth. The second chapter reveals roles that humans play in interaction with technical systems and provides an overview of methods used in technology to take into account human factors and manage them in dependence on these roles. In the third chapter, the focus shifts to human role when creating a technical system, in the design stage. The existing approaches to human factors overviewed in the second chapter are analyzed from the standpoint of designer. The analysis makes it possible to obtain a generalized assessment of these approaches and identify flaws in the safety-ensuring framework, which is generally reliable. In the fourth chapter, appropriate changes in the design paradigm, design principles and methods are proposed to eliminate the identified flaws and implement advances views on safe human–machine interaction. The topic of the fifth chapter is long-lived technical systems, including sociotechnical ones, in which specific aspects of human factors are manifested, and due to their evolution, additional challenges for design arise.

This book is intended primarily for designers, engineers and researchers, as well as for heads of design and engineering organizations working in aviation and space engineering, nuclear power or creating other energy-saturated and therefore potentially dangerous machines and infrastructure facilities. The author hopes that the presented theoretical provisions and practical examples will help specialists in understanding the problems associated with human participation in the life cycle of such systems. On this way, they will be able to benefit from the experience of not one, but a number of industries.

The book can be useful also for students of technical universities and graduate students as it contains a systematic overview of some design issues not addressed in other monographs, information about the current state of research and development in the field of safety and extensive bibliography. To perceive certain issues, knowledge is required in the scope of courses on the theory of elasticity and plasticity, structural mechanics, reliability theory, automatic control theory, as well as familiarity with computational methods of applied mathematics, mechanics and optimization. Along with the professional and student auditorium, factual data on achievements in hi-tech industries in the USA, Russia and European countries can attract the attention of non-specialists—those who are not indifferent to engineering science and follow its development, i.e., technology-fascinated readers.

Chapter 2
Existing Approaches to Human Factors in Technology

2.1 Human Roles in Interaction with Technical Systems

In any technical, human–machine system, human is a specific constituent part, which in principle differs from other components: structure, engine or power unit, control devices, etc. The machine elements are characterized by sufficiently defined properties and unambiguous subordination to physical laws. Human is a fuzzy part whose behavior depends not only on physical effects, but also on his or her decisions made on the basis of information, following insufficiently known rules.

The existing approaches to accounting for this phenomenon can be divided depending on the role of human in interaction with a technical system. As the result of this dividing, two significant roles are distinguished: the human controlling the system and accordingly its operation, and the human subject to control (an object to be managed) during the operation. The first role, which is designated as *operator*, is performed by an airplane pilot, a ground vehicle driver, an air traffic controller, etc. A more general category that characterizes the role of this kind is *personnel.* Personnel include, in addition to operators, other humans ensuring the operation of the system: service engineers and maintenance technicians, security staff, heads of relevant departments, etc.

A significant part of the systems controlled and maintained by personnel are in contact with other "bearers" of human factors: passengers transported by air or ground vehicles, population living nearby to nuclear power plants, and so on. Some technical systems are not controlled directly by humans, but function in the human habitat—they include, for example, unmanned vehicles, robotic production facilities, as well as buildings, bridges and other civil infrastructure objects used by consumers of services. Many of these systems are potentially dangerous due to the presence of flammable, toxic or radioactive substances in them, the saturation of energy that could release in an uncontrolled way during failure or destruction, or due to difficult evacuation in case of an emergency, etc. The second role of human in interaction with technical system, being not personnel (which controls) but an object of control

Y. Spirochkin, *Human Factors and Design*,
https://doi.org/10.1007/978-981-19-8832-5_2

or management—passengers, members of the population living in certain territory or consumers of services provided by the system, can be designated by the term *the public*.

The difference between these two roles is determined by the nature of human interaction with the system, and the fuzziness of human factors has an additional effect. Personnel members must perform their duties prescribed in regulations, manuals and job instructions. The public is supposed to obey directions and notices of personnel, but it does so only to the best of its understanding and desire; there is sufficient randomness in its behavior. In either of both categories of roles, there can be mentally unbalanced individuals and intruders. In threatening circumstances, the public can turn into a crowd, and its behavior can become difficult to control. Emergencies are a threat to the public, but they themselves can be caused by negative manifestations of human factors from the public side: irresponsibility, panic, aggression, etc. The most extreme of these manifestations is terrorism.

It should be noted that the boundary between the roles *personnel* and *the public* is not immovable: for example, pilots of a spacecraft during its launch into orbit or when entering the Earth's atmosphere are actually passengers, since they are not able to control the system in the corresponding extreme physical conditions. They become *personnel* only in orbit or in the final phase of reentry—landing, when they can perform the functions of operators of various systems. On the other hand, terrorists from the public who hijack an airplane or a building take over the control of the system for their own purposes (assuming the role of operator).

In emergency situations, human factors can manifest themselves in specific human roles—when humans, whether personnel or the public, become "hostages" of the events that develop without possibility to control them, i.e., "on their own". Such situations are, for example, crashes of fast-moving vehicles including aircraft or helicopter crash landings, collisions of cars or trains with obstacles or between themselves, etc. Operator control in the events of this type is impossible due to the high speed and short duration of the physical processes taking place. The corresponding human role can be designated as *vehicle occupants in crashes*. The interest to analyze human factors in this instance is determined by the need to find methods to ensure survival under relevant specific conditions and implement them.

Emergencies of another kind are created by malicious actions, including terrorist attacks. The character of these situations also limits the human ability to control them. Threats to human life in such cases are not due to natural effects or the consequences of technical equipment failures, which are expected to some extent, but are caused by unpredictable manifestation of human factors in one or more intruders who use the technical system they control for their own purposes. The situations of this kind include not only terrorist acts, like the attack on the World Trade Center in New York in 2001 mentioned in Sect. 1.2, but also the crashes of airplanes or other vehicles due to deliberate actions of crewmembers. Such actions were committed, for example, by

- the captain of the Embraer 190 aircraft of *Linhas Aéreas de Moçambique* (LAM), Erminho dos Santos Fernandes, who under the influence of emotional shocks

purposely sent the aircraft to collide with the ground on November 29, 2013; as a result of it, 27 passengers and 6 crewmembers were killed, or
- the co-pilot of *Germanwings* Andreas Lubitz, who was suffering for a long time before from a mental illness, and intentionally killed himself and 149 other people on March 24, 2015, by directing the A320 plane, in the temporary absence of the captain in the cockpit, into a mountain.

A complete list of the deliberately organized aviation accidents can be found in (ASN 2015).

The behavior of those who play the corresponding dramatic role: *people under conditions of malicious actions* can be described using two sub-roles: *attacker* and *potential victim.*[1] The first sub-role deserves attention as a specific cause of impacts on technical systems and through them on potential victims. Scrutiny of the second sub-role is due to the need to ensure survival of people in such emergency situations.

The roles and sub-roles listed above and the features of human behavior corresponding to them, which are not governed by known physical laws, make it difficult to describe human factors in order to ensure safety. A large number of studies have been carried out in this area, wide spectrum of models approximating features important for safety has been developed, and many articles and monographs have been published. The most significant results were transformed into provisions of regulatory documents (rules, codes, standards, etc.). Before we begin to consider them, it should be noted that the specificity of some branches of technology and sensitivity of their connection with safety are manifested themselves, among other things, in restrictions of information. Published sources may not contain data on the latest approaches to human factors. Therefore, the overview below does not claim to be complete, but it serves only to demonstrate the range and main components of the existing methodology for accounting for human interaction with technical systems and controlling this interaction.

2.2 Personnel

2.2.1 Characteristics of Human Operator

Human–machine interaction has become the subject of intensive studies in the 1950s (see Sect. 1.1)—primarily because of the needs of military and then civil aviation. Additional impetus to research and development (R&D) in this area has been given in the 1960s in connection with the creation of manned spacecraft and origination of nuclear power. The results of this R&D were included in regulatory documents and relevant manuals.

Human–machine interaction at the biomechanical level is considered, e.g., in the corresponding section of the reference book Klyuev (1982). It presents the testing

[1] "Potential" means that there is a chance of survival.

methods and equipment used to determine the functional capabilities of human operator and provides experimental data on the effects of vibration, pulse accelerations, and acoustic loads on human body connected with the operation of machines, including the limits of tolerability to these loads (tolerance limits).

Vibration can lead to difficulty in breathing, deterioration of visual function and coordination of movements, and a general decrease in the quality of the operator's work. Its effect is determined by the intensity and spectral composition of vibration process, direction and duration of the loading. The indicators of intensity are average square values and amplitudes of vibration accelerations, vibration velocities, or vibration displacements at the operator's workplace, as well as logarithmic levels of these parameters in decibels (dB). The human response to vibration is assessed based on the dynamic characteristics that characterize oscillations of his or her body: natural frequencies, transfer functions, etc. Resonant oscillations of individual parts and organs of the body occur in the following frequency ranges (Klyuev 1982):

- head—8 to 27 Hz;
- eyes—12 to 27 Hz;
- hands and feet—2 to 8 Hz;
- abdomen—4 to 12 Hz, etc.

These and other dynamic characteristics are used as input data when designing workplaces, vibration protection systems and for development of hygienic standards taking into account vibration effects.

Mechanical impacts and other pulse loadings in human–machine systems (associated, for example, with ejection of pilots, landing of aircraft, collision of ground vehicles) are characterized by linear accelerations, measured in units of g (free fall acceleration, equal to 9.81 m/s^2 at ground level). Pulse acceleration as function of time t can be approximated by a trapezoid, and three parameters are used to determine the human tolerability to such pulse: the maximum value of acceleration, $a_{\max}$, the rate of increase of acceleration, $\frac{da}{dt}$, and the pulse duration τ (see Fig. 2.1). Sometimes, triangular pulses are considered (in which interval $t_1 - t_2$ is zero).

Allowable parameters of acceleration pulses for human operator (corresponding human tolerance limits), established on basis of experimental data and taken from the book Klyuev (1982), are shown in Table 2.1.

The data in Table 2.1 are conditional because they relate to certain combinations of all three parameters. If the value of any parameter would be changed, the values of the remaining ones should change accordingly. The presented data do not contain human tolerance limits for lateral accelerations among. They can be found, for example, in the paper Shanahan (2004) referring to the design guide (USAAVSCOM TR 89-D-22B 1989). According to these sources, the allowable value of lateral accelerations is 20 g when the pulse duration is 0.1 s.

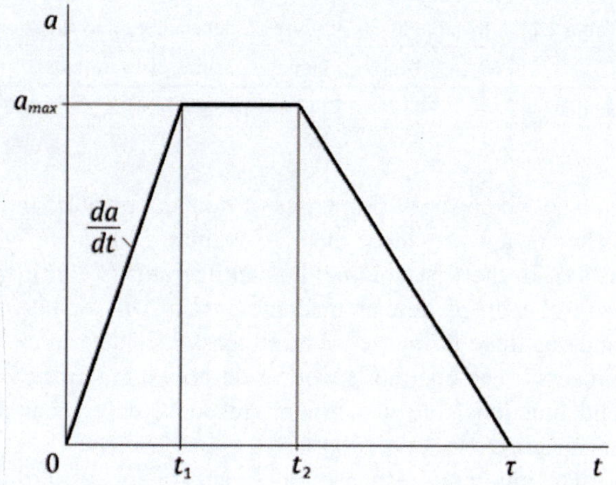

Fig. 2.1 Pulse acceleration

Table 2.1 Allowable parameters of acceleration pulses for human operator fixed in the seat with shoulder and waist belts (Klyuev 1982)

Direction of acceleration	Maximum value of acceleration ($a_{\max}$), g	Rate of increase of acceleration ($\frac{da}{dt}$), g/s	Pulse duration (τ), s
Head to pelvis	20	300	0.5
Pelvis to head	7	300	0.1
Chest to back and back to chest	40	500	0.5

The main means of protecting of humans from such impacts are shock-absorbing seats, restraining devices (belts and airbags) and protective helmets.

Acoustic effects on humans (noise during machine operation) are evaluated using the following indicators:

- sound pressure level, dB A[2];
- level of the main (first) harmonic characterizing the low-frequency part of the internal noise spectrum, dB A;
- articulation index characterizing the high-frequency part of the noise spectrum and, accordingly, the ability of human to conduct a conversation perceived by an interlocutor;
- limiting noise spectra.

These indicators make it possible to assess the acoustic comfort in the workplace. The maximum allowable levels of noise exposure depending on its duration (which also affects the operator's state) are shown in Table 2.2.

[2] *A* means the frequency response of a noise meter, which is consistent with the perception of sound by human.

Table 2.2 Allowable levels of noise[3] L depending on its duration T (Klyuev 1982)

T	30 s	1 min	2 min	7 min	15 min	0.5 h	1 h	2 h	4 h	8 h
L, dB A	120	117	114	108	105	102	99	96	93	90

The whole set of properties of human operator, and relevant features of human behavior are presented in the 6-volume *Handbook of Human Factors* (Salvendy 1987). In the first volume, human operator is considered as a specific link in the control loop of human–machine system. His or her psychomotor, informational, and cognitive properties, as well as skills, "determine content of the entire control process". The operator's work is described in terms of cybernetics, using structural and functional block diagrams, response delay characteristics, signal bandwidth, feedback, etc.

The minimum response times vary for signals of different sensory modalities in the range from 0.14 s (hearing) to 0.52 s (body rotation). The average delay time of human reaction is about 0.3 s, and this value is one of the main limitations when making decisions by the operator.

The operator's bandwidth takes up a range approximately up to 1 Hz. If the operator has to track sinusoidal oscillations of a higher frequency, then usually the quality of tracking deteriorates. But if the input signal is predictable, the operator can overcome this limitation and with a necessary training expand the bandwidth up to 2 Hz.

An important factor in the operator's work is nonlinearity of the control system. When considering the human–machine interaction in the coordinates shown in Fig. 2.2 (system response vs. human operator's control action), any deviation from the linear function (1) worsens the controllability. Nonlinearity requires the operator to anticipate effect of a control action in a more complex way and, accordingly, to have special properties when implementing feedback. The worst type of nonlinearity is that presented by curve 3.

Errors of human operator are usually characterized by the probability of their occurrence, which is defined as "the ratio of number of errors committed to number of possible errors" (Salvendy 1987), volume 1. This indicator is very arbitrary due to uncertainty of the denominator—it is "difficult to determine, since error possibilities may be hidden". In relation to human operator, the term "reliability" means the "probability of a correctly or satisfactorily completed task". Errors and reliability of human operator are analyzed based on the tree of events and probabilities of possible outcomes of control tasks.

Volumes 2, 4 and 5 of the handbook (Salvendy 1987) provide an ergonomic basis for designing human–machine systems. Volume 3 focuses on operator activity modeling, professional training and selection. In particular, it presents the transfer functions of human operator in compensating control systems, which are applicable in

[3] Equivalent values taking into account the distribution of sound pressure over frequencies in the range from 31.5 to 8000 Hz.

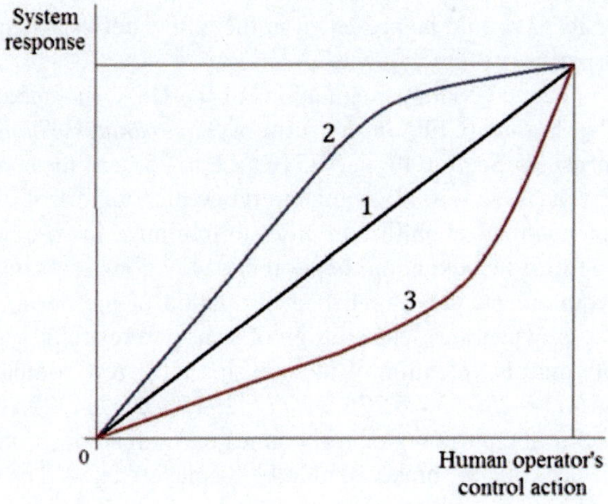

Fig. 2.2 Types of nonlinearity of the control system: deviations (2) and (3) from the linear function (1)

vehicle control tasks. Volume 6 discusses the features of human–machine interaction in automated systems, including computer systems.

2.2.2 Regulation in Relation to Personnel Performance

The results of human–machine interaction studies led to the development of regulatory requirements to take into account the influence of personnel on the functioning of technical systems and to maintain the personnel performance in order to ensure safe operation. Relevant requirements concerning NPP personnel are given, for example, in Russian document of the FNP category (NP-001-15):

- NPP design "shall provide for means by which errors of personnel are prevented or their consequences are mitigated, including errors made during maintenance or repair";
- each NPP unit must be equipped with an operator information support system.

Another document of the FNP category, (NP-003-97), includes requirements for technical means of training plant operators. One of the safety guidelines being in use in Russian nuclear power industry, (RB-100-15),[4] contains recommendations for performing reliability analysis, and the probabilistic characteristics of erroneous human actions. Recently published document (RB-166-20) recommends designers to account for the uncertainties caused by errors of NPP personnel when compiling lists of design-basis initial events and beyond-design-basis accidents. Taking into account the absence in regulatory documents of any established distribution functions and ranges of parameters characterizing the actions of personnel, the choice of data for

[4] The designation of these guidelines includes RB (in Russian "РБ"), the guideline number and two digits indicating the year of issue.

analysis should be presented in the NPP safety substantiation report "including the use of expert assessments".

Federal Aviation Regulations of the USA, included now in the *Code of Federal Regulations* (CFR) in the form of *Airworthiness Standards* for transport category airplanes, Section 14, Part 25 (CFR Part 25), and the corresponding Russian *Aviation Rules* (AP-25 2009) regulate ergonomic support of aircraft crews as a means of maintaining reliability. In order to minimize interference with piloting, which can cause erroneous actions of pilots and violation of safe operating conditions of aircraft, requirements are established in relation of the design of cockpit, arrangement of the crewmember seats, design of seats and restraint systems, placement and design of controls, retention of items of mass in crew compartments (§§ 25.671, 25.672, 25.771—25.773, 25.775, 25.777, 25.779, 25.781, 25.785, 25.787, 25.789).[5]

For the purposes of ergonomics and safety, positions of all wing mechanization devices and air brakes should be displayed, as well as emergency alarms about the configuration of aircraft unacceptable for safe take-off or landing, warnings about ground proximity generated by appropriate automatic systems (§§ 25.699, 25.703, 25.729, 25F.8.2, 8.2.2.12). In accordance with § 25F.8.2 (8.2.2.7), aircraft must be equipped with autopilot, trajectory control systems, takeoff thrust control system, etc. Their function is aimed at facilitating the work of pilots and minimizing piloting errors.

Provisions on ergonomic support for the operators of stationary and mobile human–machine systems are established in Russian standards of the following series:

- *Man–machine system*, including Soviet-era standards (GOST 21958-76), (GOST 22269-76), and (GOST 23000-78)[6];
- *Standards system of ergonomic requirements and ergonomic means*, combining the relevant Soviet and post-Soviet standards (GOST 29.05.002-82), (GOST R 29.05.008-96), (GOST R 29.08.004-96),[7] etc.;
- *Ergonomics*, which covers national standards (GOST R ISO 26800-2013), (GOST R ISO 6385-2016), (GOST R ISO 7250-1-2013), (GOST R 56620.2-2015) and (GOST R ISO 7250-3-2019) reproducing the appropriate international standards developed at the International Organization for Standardization (ISO);
- *Ergonomics of human–system interaction*—this series also reproduces some ISO standards and includes (GOST R 55241.1-2012), (GOST R ISO 9241-161-2016), etc.;
- *Ergonomics of vehicles*—a similar "remake" presented by (GOST R 55237.1-2012) and two other standards with the same generic index.

[5] The numbers of paragraph here and below correspond to the Russian *Aviation Rules* (AP-25 2009).

[6] GOST is an abbreviation of the Russian term "Gosudarstvennyi Obschesoyuznyi STandart" that means *State all-union standard*. This designation was used for standards in the former USSR and then goes in the use by tradition for Russian standards. Two or four digits after the numerical designation of the standard and the dash indicate the year of its issue.

[7] The abbreviation GOST R is used for national standards of the Russian Federation.

Relevant requirements and rules are given in many other documents on standardization, among them (GOST 20296-2014), (GOST R 56274-2014), (GOST R ISO 11064-1-2015), (GOST R ISO 11064-2-2015).

Along with the regulation of ergonomics, a system of occupational safety standards is used in the Russian Federation, named in Russian: "Sistema Standartov Bezopasnosti Truda" (SSBT). It covers "requirements, norms and rules aimed at ensuring safety, preservation of health and working capacity of human in the process of work, except for issues regulated by labor legislation»" (GOST R 12.0.001-2013). SSBT is divided into groups of documents that contain: organizational and methodological provisions (group 0[8]), requirements and norms for hazardous and harmful production factors (group 1), safety requirements for production machinery and equipment (group 2), production processes (group 3), and protective means (group 4).

Requirements of ergonomic support apply also to the working processes of maintenance and repair of technical systems, as well as to the information tools used in these processes (hardware and software). Military aviation has become the first area, in which, for example, interactive electronic technical manuals (IETMs) began to be used. Currently, the Russian national standard (GOST R 54088-2017) regulates their use in all industries.

2.2.3 Rescue in Emergency Situations

Extreme operating conditions, characteristic of some types of human–machine systems, require ensuring the survival of personnel in possible accidents that can get out of control and develop into catastrophe.

For example, in order to rescue pilots of military aircraft, ejection seats are used. Such a seat is pushed out of the plane by an explosive charge or a solid-fuel rocket engine, taking the sitting human with it. Then the parachute opens and the landing takes place. When ejecting, pilot experiences acceleration of approximately 12–14 g. In the 1960s and 1970s, accelerations acting on the pilots of Soviet military planes in such emergency situations reached 20–22 g. A side effect of ejection often was vertebral fractures.

The developer of ejection seats was in the Soviet Union and remains the same in Russia the Scientific and Production Enterprise (in Russian, it sounds as "Nauchno-Proizvodstvennoe Predpriyatie") *Zvezda.*[9] Figure 2.3 shows one of its products—the ejection seat K-36D-5 designed for installation in high-speed maneuverable planes of the latest generation including T-50 or SU-35S. This seat provides rescue of a crewmember in the range of aircraft velocities up to 1300 km/h at altitude from ground

[8] Group number is the second digit of the numerical designation of the standard, for example, GOST 12.1.025-2012, GOST 12.3.036-2012, etc.

[9] The full current name of this enterprise is JSC Academician G. I. Severin Scientific and Production Enterprise *Zvezda*.

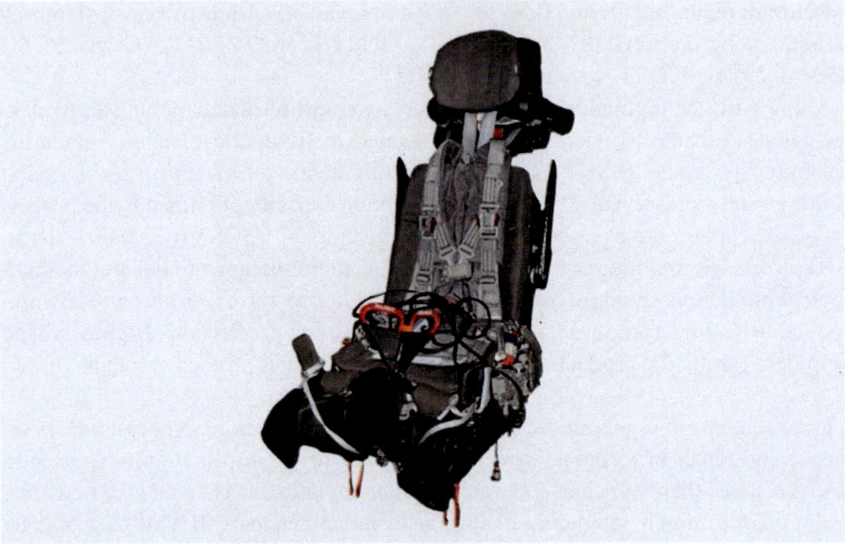

Fig. 2.3 The ejection seat K36 D-5. Reproduced from http://www.zvezda-npp.ru/ru/node/118 with the permission of the Scientific and Production Enterprise *Zvezda*

level to 20 km and is used with a set of protective equipment and oxygen equipment. K-52 military helicopters are equipped with seats KAS-37-800 M, which provide not only the safe emergency escape of the helicopter by ejection, but also a reduction of acceleration during impact on the ground to allowable values. Helicopters of this type were the first in which the ejection systems were used. Their main propellers are equipped with pyrobolts in order to reset the blades just before the ejection of seats.

Ejection seats have been installed in the Soviet spacecraft *Vostok* and the American spacecraft *Gemini*, as well as the space shuttles *Columbia* and *Enterprise* before their first flights with two crewmembers. Such seats were planned to be installed in the Soviet *Buran*, but this was never implemented. The only civilian jet with ejection seats was the Soviet supersonic plane Tu-144—more precisely, they were installed only in its prototype and they were intended only for crewmembers, not for passengers.

In some aircraft types, a concept of crew escape capsule was used instead of ejection seats. For example, in the US military aircraft F-111, it was possible to eject the airframe section containing the crew cabin (see Fig. 2.4). In abnormal situation, solid-fuel rocket engines could separate it from the aircraft and take it away. After the parachute descent of the rescue capsule, its soft landing was provided by a shock absorption system using an airbag (see Sect. 2.4.4). The latter also played the role of flotation equipment in case of landing on water.

Another embodiment of this concept is the Launch Escape System (LES) designed to rescue a crew of spacecraft during its launch, if an emergency situation occurs in the time interval from being on the launch pad almost to entering the orbit. The

Fig. 2.4 The crew escape capsule of F-111. *Photo* Greg Goebel. Reproduced from https://www.flickr.com/photos/37467370@N08/7331359342. CC BY-SA 2.0

LES includes small solid-fuel engines mounted directly on the spacecraft (American projects *Mercury*, *Apollo*) or on the head fairing (Russian project *Soyuz*).

The system used on *Soyuz* is named in Russian "Sistema Avariinogo Spaseniya" (SAS) that means emergency rescue system (see Fig. 2.5). In emergency situation, when the launch is interrupted, SAS is able to detach the spacecraft with the crew from the launch vehicle rocket, take it away and provide a soft landing using the parachutes and special solid-fuel brake engines. This system is activated before the launch starts and can be triggered automatically when a launch vehicle rocket failure is detected or by manual command. During the work of SAS, crewmembers can be exposed to high accelerations, up to 20–25 g.

The first emergency rescue of cosmonauts by SAS, one of few such cases, has occurred on September 26, 1983, during an accident of the *Soyuz* launch vehicle on the launch pad. Systems of this kind are now used in all manned spacecraft being in operation, under design or testing, which have a relatively simple, ballistic-type configuration and structure. It is believed that neither ejection seats nor escape capsules are effective for more complex aerospace vehicles, for example, space shuttles (NASA Wiki 2016b). The author, like some other specialists, is of opposite opinion and develops it in Chap. 4.

For emergency situations which are possible during the work of personnel of orbital stations, designers of the enterprise *Zvezda* developed the Cosmonaut Rescue Aid (it is named in Russian "Ustanovka spaceniya kosmonavta", USK) to be installed on the spacesuit *Orlan-M* (see Fig. 2.6). USK is a satchel-type device set with reactive

Fig. 2.5 The SAS installed on the spacecraft *Soyuz-TMA*. *Photo* NASA. Reproduced from https://wiki2.org/ru/%D0%A4%D0%B0%D0%B9%D0%BB:Jsc2006e11326_jpg. In the public domain

gas micro-engines designed for maneuvering in support-free space in order to return a cosmonaut who has lost contact with the surface of the station. USK was provided for the crews of the Russian orbital station *Mir*; such rescue aid sets are not used at the International Space Station (ISS).[10]

In the US Navy, the following means are in service to evacuate crews from submarines in distress:

- at depth up to 180 m—individual rescue equipment SEIE (Submarine Escape and Immersion Equipment);
- at depth up to 260 m—underwater rescue chambers SRC (Submarine Rescue Chamber), each of which is controlled by two operators and is able to lift 6 submariners.

The most modern development, Submarine Rescue Diving and Recompression System (SRDRS), is currently in the implementation stage (Armscom 2017). This system provides evacuation from depth of up to 610 m. It consists of the rescue module Pressurized Rescue Module (PRM), two decompression pressure chambers Submarine Decompression System (SDS) that are transported on a carrier ship and are connected to PRM after its surfacing, descent and lifting devices, diving suits, control and navigation equipment, etc.

[10] According to information from Dr. Oleg A. Saprykin.

Fig. 2.6 The space suit *Orlan-M* equipped with USK. Reproduced from http://www.zvezda-npp.ru/ru/node/76 with the permission of the Scientific and Production Enterprise *Zvezda*

The *International Convention on Maritime Search and Rescue* (SAR)[11] establishes obligations of the participating countries (including the Russian Federation) to provide assistance to people in distress at sea, as well as in airspace above it. Measures implemented in accordance with these obligations shall apply to any person in distress, including crews of ships and vessels, airplanes and offshore facilities (e.g., offshore mining platforms). The conditions for the evacuation of people from ships in emergency situations, the actions of personnel during evacuation, and the requirements for rescue equipment are regulated by the *International Convention for the Safety of Life at Sea* (SOLAS 1974). Structural parameters of lifebuoys and vests, wetsuits, lifeboats, and other relevant equipment are established by the *International Life-Saving Appliance Code* (LSA Code 2017). Part II of the Russian *Rules for the Equipment of Sea Vessels* (RMRS 2022) contains detailed requirements for vessels built after July 1, 1998, with regard to their providing with life-saving equipment and devices, as well as characteristics of such equipment and devices. Actions for the search and rescue of people in distress, including crewmembers of ships and aircraft, are provided for in Chapter XIII of the *Air Code* of the Russian Federation.[12]

To date, organizational and technical approaches are developed to rescue the personnel of offshore oil or gas production platforms in emergency situations. They can be found, for example, in the publication (CAAP 71 2016). The main method of rescue is the evacuation of people by helicopter from the helicopter deck of the

[11] SAR has been adopted on April 27, 1979 and is effective since June 22, 1985. Last amendments to it have been adopted on May 2004 and are effective since July 1, 2006.

[12] Federal Law of the Russian Federation (as amended on June 11, 2021).

platform (helidesk). However, evacuation of this kind can cause additional problems. Chapter 18 of the reviewed publication includes a manual for emergency response in all helicopter-related abnormal events. The range of such events covers: helicopter crash on the helideck or in the neighborhood (with immersion in water), wrong deck landing, emergency medical care by helicopter, fuel spillage, fire during helicopter refueling operations, helicopter use for man over-board, etc. The purpose of emergency response and appropriate planning is to save lives and ensure the operation of helicopters. Helideck personnel must be trained to perform their duties and have the necessary competence to meet the challenges that arise. Chapter 19 of the publication (CAAP 71 2016) describes the requirements for personnel and their training programs.

In Russia, the national standard (GOST R 58217-2018) is used for planning, organizing, and implementing rescue measures, which takes into account the Arctic conditions of oil and gas production.

2.2.4 Applicable Models

An information model SHELL is widely used in aviation for analyzing the interaction of human with other parts of human–machine systems. This model, proposed by Frank Hawkins for such analysis (Hawkins 1993), is a logical structure composed from the following elements (see Fig. 2.7):

- S (Software);
- H (Hardware);
- E (Environment);
- L (Liveware).

In this context, "Software" designates non-physical, intangible aspects of the aviation system governing its operation and organization of information within it: computer programs, aviation regulations, rules, standards, instructions, procedures, practices, formal conventions, habits, etc.

"Hardware" covers physical elements of the system, such as aircraft structure, equipment, mechanical and electronic devices, controls, control surfaces, operator tools, computers, safety appliances and so on.

Fig. 2.7 The SHELL model

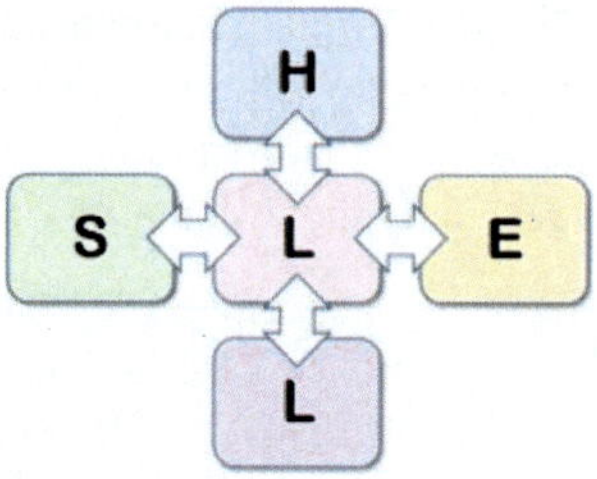

"Environment" is the conditions in which the system operates. These conditions can be divided into internal and external ones. The first category (internal environment) relates to immediate work area and includes physical factors, e.g., the temperature in the cockpit, air pressure, humidity, noise, vibration, ambient light levels. The second category includes the physical environment outside the immediate work area: weather, terrain, airspace, facilities and infrastructure including airports, as well as in a broad sense organizational, economic, regulatory, political and social factors that may affect the work of personnel.

"Liveware" means people involved in the operation of the system: flight crews, air traffic controllers, ground maintenance teams, managing personnel, security staff and others. This element is introduced to consider human performance, capabilities and limitations. The use of two such elements allows analysts to take into account the influence of one part of the personnel on the other.

Possibilities to minimize errors associated with human factors that can lead to adverse outcomes have been investigated by James Reason. He has proposed an information model visualizing the causation of accidents or catastrophes: the Swiss cheese model (Reason 2000). According to this model, safety barriers and levels of protection in human–machine system (generically called "defensive layers" or "defenses") are like slices of Swiss cheese containing many "holes" (see Fig. 2.8). The locations and sizes of these "holes" are not obvious and they can change during the life cycle of the system. Defensive layers can be overcome when "holes" line up at some point in time to provide for a "trajectory" of the possibility of an accident leading from "Hazards" to "Losses". There are two reasons for the "holes" to arise: active failures (unsafe actions of humans or malfunctions of material components) and latent conditions (including design or building defects and hidden errors).

Fig. 2.8 The Swiss cheese model

The article (Reason 2000) indicates (with reference to a study carried out in the USA) the signs of high reliability systems—organizations operating in hazardous conditions and having a share of accidents that is less than statistically expected. These organizations include US Navy nuclear aircraft carriers, nuclear power plants and air traffic control centers. They correspond to the concept of "resilient system"—a system which resists adverse external impacts due to its flexibility, can reconfigure itself to suit local circumstances and recover after impacts.

Both models described above are included in the Safety Management Manual of the International Civil Aviation Organization (ICAO). A significant part of its latest, Fourth edition (ICAO 2018) is devoted to the consideration of human factors and how to manage them at all levels of aircraft operation support.

2.2.5 Personnel Management for Safety Purposes

Approaches to ensuring safety through human factors management, that provide for improving the quality of work and the degree of responsibility of personnel, their ability to anticipate hazards and respond correctly to them, have now taken shape into three interrelated directions:

(1) Safety culture,
(2) Safety II (safety of the second level) and
(3) Resilience engineering (creating sustainable, self-recovering systems).

The importance of the first of these directions is recognized by international organizations dealing with safety issues, and national regulatory authorities. The term *safety culture* appeared probably for the first time in the ICAO Human Factors Training Manual (ICAO 1998). It was widely used in the document supplemented this manual (ICAO 2003), which described the principles of human factors accounting and management in aircraft maintenance. Safety culture is the content of Chapter 3 of the manual (ICAO 2018), mentioned above. According to this manual, safety culture implies:

- awareness of the management and employees in an organization of the risks and known hazards;
- continuous activities to preserve and enhance safety;
- access to the resources required for safe operations;
- readiness and ability to adapt when faced safety issues;
- informing about safety issues;
- consistent assessing the safety related behaviors throughout the organization.

The basic document of the IAEA Safety Standards Series, SF-1 (IAEA 2006a), considers the formation of safety culture as an integral element of the management system at each nuclear facility. In the context of nuclear and radiation safety, this culture encompasses:

- individual and collective commitment to safety on the part of the management and personnel at all levels;
- accountability of organizations and individuals at all levels for safety;
- measures to encourage a questioning and learning attitude with regard to safety;
- systematic assessments of safety indicators;
- the feedback and analysis of operating experience, so that lessons may be learned, shared and acted upon.

In accordance with the Russian document of the FNP category (NP-001-15), safety of nuclear power plants is achieved, along with technical solutions, through the "formation and support of safety culture", as well as aspects related to it and dependent on human factors: learning from the operating experience and continuous development of science and technology. Safety culture does not belong to quantitative indicators (describing the technical perfection of NPP elements or the stability of physical processes in them). It covers "a set of characteristics and features of activities of organizations and behavior of individuals, which establishes that the issues of NPP safety, as having the highest priority, should be given attention determined by their significance" (NP-001-15).

Aspects of safety culture in space flight are regulated to a lesser extent; they can be found, e.g., in the book Sgobba et al. (2018).

The terms *Safety* II and *Resilience engineering* and the corresponding directions of organizational work on human factors management aimed at improving safety appeared at the turn of the 2000s. The following circumstances contributed to the formation of these directions:

(1) understanding of safety in the previous period of time as "the absence of accidents and catastrophes", which actually made it difficult to develop a rational approach to achieve this goal;
(2) increasing the reliability of equipment, skills and responsibility of personnel—traditional approaches to safety, designated as Safety I, which have reduced the probability of accidents and catastrophes for modern technical systems to a very small level, practically not subject to experimental registration and control (see Sect. 1.2);
(3) the remaining possibility of extremely unlikely dangerous events for which there are no statistics, that makes it impossible to predict them and manage the risk in the frame of Safety I.

Highly unlikely and unpredictable events covered by the *Black Swan* metaphor (Taleb 2007) present a problem that defies experimental investigation. The need to manage "what is not" (the vanishingly low probability of dangerous events in the face of a lack of statistically sound data or their complete absence) forces organizations and individuals to move away from traditional ways of ensuring safety. Management is shifting to an area of creating and maintaining the necessary flexibility and adaptability of personnel (and, accordingly, the processes they perform) in changing, sometimes unpredictable working conditions (Hollnagel 2018). If the traditional approaches to ensuring safety, Safety I, can be named "execution safety", then for Safety II, the characteristic "readiness safety" is suitable. Thus, two-level protection against accidents and catastrophes is constructed, in which the second level (Safety II) refers mainly to the organizational policy for creating and developing the potentials of flexibility and adaptability (resilience potentials) of personnel.

The direction of Resilience engineering has a broader scope of application; it also covers technical artifacts created by human.

2.3 The Public

2.3.1 The Public as a "Passive" Object of Management

There are two main approaches to taking into account human factors embodied in the public. The first of them, typical for regulatory documents, considers humans as a "target" of possible hazardous impacts; humans shall be protected or evacuated out of the affected areas if the allowable limits of such impacts are exceeded. This approach can be described as treating the public as a *"passive" object of management* in emergency situations. In Russian nuclear power industry, it is presented, for example, in the following documents[13]:

- regarding the allowable levels of radiation exposure[14]—in (SanPiN 2.6.1.2523-09);
- in relation to the requirements for the placement of nuclear power plants, their design and operation in accordance with the radiation safety criterion—in (SanPiN 2.6.1.24-03) and (SP 151.13330.2012);
- in the area of fire safety during operation of nuclear power plants—in (STO 1.1.1.04.001.1500-2018);
- in the area of civil defense measures—in (SP 165.1325800.2014).

Airworthiness Standards of the USA (CFR Part 25) and Russian *Aviation Rules* (AP-25 2009) contain provisions for ensuring safety of aircraft passengers in various flight conditions. For normal operation, these provisions apply to the opening and closing of fuselage doors (§ 25.783), seats, berths, and safety belts (§ 25.785), design of stowage compartments (§ 25.787), retention of items of mass (§ 25.789), passenger information signs and placards (§ 25.791), condition of floor surface (§ 25.793) and so on. For cases of emergency landing or ditching, the following parameters are specified:

(1) allowable values of inertial forces acting on passengers (as well as crewmembers);
(2) required strength of seats and seat belts;

[13] The list is not exhaustive. The year of issue of each document is indicated by the last two or four digits.

[14] For the population and personnel.

(3) location and arrangement of emergency exits;
(4) width of aisles and characteristics of auxiliary means of leaving the aircraft by passengers (and crewmembers);
(5) emergency evacuation time limit.

The first two positions relate more to the other role category of human factors, considered in Sect. 2.4. The others take into account the possibility of fire due to destruction of aircraft during an emergency landing on ground, or immersion in water when ditching. Based on these risks, the evacuation time should not exceed 90 s (§ 25.803).

The Russian *Air Code* provides, as already mentioned above (see Sect. 2.2.3), for actions to search air vessels in distress and to rescue people who are (were) in them. Similar measures in relation to the passengers of sea vessels are established in the documents mentioned in Sect. 2.2.3: *The International Convention on Maritime Search and Rescue*, *The International Convention for the Safety of Life at Sea* (SOLAS 1974), *The International Life-Saving Appliance Code* (LSA Code 2017) and the Russian *Rules for the Equipment of Sea Vessels* (RMRS 2022).

General provisions on ensuring safety of the public in emergency situations are contained in the Federal laws of the Russian Federation:

- *On the protection of the population and territories against natural and man-made emergencies*[15];
- *On emergency rescue services and the status of rescuers*[16];
- as well as in a number of other regulatory legal documents and standards of GOST R 22 series: (GOST R 22.0.02-2016), (GOST R 22.0.03-95), (GOST R 22.3.01-94), etc.

Safety of the public provided by buildings and infrastructure is regulated by the Federal laws: *Technical rule on safety of buildings and facilities*,[17] *Technical rule on fire safety requirements*,[18] relevant codes of rules, for example, (SP 1.13130.2009), (SP 3.13130.2009), (SP 117.13330.2011), (SP 255.1325800.2016), (SP 309.1235800.2017) and standards specified in the approved *Lists of documents in the field of standardization*.[19]

[15] The Federal Law adopted by the State Duma on November 11, 1994, as amended on December 8, 2020.

[16] The Federal Law adopted by the State Duma on July 14, 1995, as amended on July 13, 2020.

[17] The Federal Law No. 384-FZ of December 30, 2009, as amended on 02.07.2013.

[18] The Federal Law No. 123-FZ of July 22, 2008, as amended on December 27, 2018.

[19] List of documents in the field of standardization, as a result of application of which on a voluntary basis, compliance with requirements of the following document is ensured: The Federal Law No. 123-FZ of July 22, 2008: *Technical rule on fire safety requirements* (Approved by the Directive No. 1317 of the Federal Agency for Technical Regulation and Metrology of June 3, 2019).

List of documents in the field of standardization, as a result of application of which on a voluntary basis, compliance with requirements of the following document is ensured: The Federal Law No. 384-FZ of December 30, 2009: *Technical rule on safety of buildings and facilities* (Approved by the Directive No. 831 of the Federal Agency for Technical Regulation and Metrology of April 17, 2019).

2.3.2 Active Subjects

The second approach to human factors, manifesting themselves in the public, is based on the assumption that people are *"active" subjects* who perform unconscious or conscious actions in the conditions of management.

The unconscious behavior of the public in the form of a crowd managed by feelings (which are excited by leaders) is considered from the point of view of psychology in the classic book by Gustave Lebon reprinted in Russia recently (Lebon 2017). First published more than a century ago, this book has become in fact a guide on human factors management, used by politicians in subsequent years. An overview of other studies, including modern ones, of mass behavior at the socio-psychological level can be found, for example, in the article Zhuravlev and Sosnin (2014). The understanding of relevant processes ongoing in crowded places allows officers of the Ministry of Internal Affairs to solve their official tasks (Shakhmatov 2012; Podlinyaev and Karimov 2018).

In order to ensure safety when the public, being a crowd of active subjects, interact with technical systems, the following aspects are of interest:

- movement of human mass (which can lead to violation of the stability of aircraft flight or the collapse of building structure);
- the possibility of rapid evacuation of people from the zone of accident or disaster, as well as obstacles created by structural elements of the system or the behavior of the evacuees themselves (in particular, their panic);
- dynamic loads acting on humans in emergency situations from the system or on the structure of the system from the part of humans, etc.

When choosing a suitable method to achieve the required level of safety, it is necessary to take into account the socio-psychological, informational and physical characteristics of human behavior, the properties of the technical system and the environment in which emergency situation develops, as well as the processes of this development. Existing methods can be classified depending on the extent of coverage of the whole complex and the type of mathematical models used.

The physical parameters of human flows (density, speed) and their changes in emergency situation (including those caused by panic) are taken into account when designing premises, buildings and infrastructure facilities of mass use, as well as other elements of urban environment (Predtechensky and Milinsky 1979). Designers use semi-empirical analytical models, including probabilistic ones, which approximate statistical data, the postulated cases of human flow formation (merger, separation, intersection) and typical scenarios for emerging and dispersing clusters of people. Application of such models is provided for by modern building codes of rules. Regulatory nature of this modeling, postulating the design cases, insignificant degree of human activity under consideration (in fact, only in the form of variation in the speed of human flow, modeled like a liquid), and neglecting other aspects but physical ones make the reviewed methodology more similar to the first approach ("passive" object of management) than to the second ("active" subjects).

Further development of this methodology is associated with the use of molecular dynamics models (based on equations of mathematical physics, therefore, it relates rather to the analytical type). In these models, in addition to physical effects, artificial forces of socio-psychological character are taken into consideration in order to describe the interaction between "molecules" that mimic humans. An overview of relevant studies is given in the article Aptukov et al. (2013), and an advanced method for modeling the behavior of panicked crowd in a closed space with complex topology is presented.

Along with analytical models of human mass behavior, modellng of acting factors and simulation of ongoing processes based on algorithms, including discrete one, is becoming increasingly widespread. Among publications in this field, books Gilbert and Troitzsch (2005) and Karpov (2005) should be mentioned: they describe a methodology for modeling collective behavior and contain examples of corresponding simulations.

In general, the development of modeling and simulation methods applicable to the analysis of collective behavior goes in the following main directions: system dynamics, discrete event simulation, cellular automata and agent-based modeling. The first direction covers the systems whose evolution over time can be expressed by algebraic, differential or finite difference equations. The methods of system dynamics were originally presented in the books Forrester (1961, 1969), but initially they were not related to analysis of the behavior of human mass. Later, fluid (continuum) and molecular dynamics models (discussed above) began to be used to describe collective behavior within the framework of these methods. But so far, these models relate mainly to collective behavior in animal groups—see, for example, the article Ouellette and Gordon (2021).

Other directions correspond to simulation of discrete events: requests in queuing systems, processes in discrete arrays of cells modeling human populations, or behavior of discrete active agents—subjects modeling individuals.

The methodology of discrete event simulation uses at least three categories of objects: customers, servers (elements performing service functions) and queues (Gilbert and Troitzsch 2005). Each of them and all together can be stochastic. A characteristic area of application of this methodology is the simulation of airport operations. Customers are, for example, passengers during the check-in and boarding processes, and servers are check-in counters and boarding gates. When passengers pass through the servers, queues arise; their length and service time are objects of minimization. This methodology is also applicable to analyzing the possibilities of evacuation of people from the airport in emergency situations.

Models in the form of cellular automata are based on the description of state of a cell depending on its neighbors. In relation to human behavior, such models are applicable to study the processes of information dissemination (Gilbert and Troitzsch 2005), propagation of epidemics (Bashabsheh 2014), etc.

Agent-based modeling can be considered the next stage in development of cellular automata. Agents are assigned the following properties (Gilbert and Troitzsch 2005):

- autonomy (relative independence from influence of others, surrounding agents);

- social communication ability (possibility to interact with each other using a specific "language");
- reactivity (understanding the environment and reacting to it);
- proactivity (ability to initiative and purposeful behavior).

An example of agent-based modeling of the behavior of human crowd in extreme situations, such as terrorist attacks, explosions or fires, is presented in the article Akopov and Beklaryan (2012). As circumstances complicating the rescue of people from a confined space, the high density of crowding and the appearing of wave effect (so-called turbulence of the crowd) are considered. This effect relates to behavior of humans when they begin to panic, push to expand their personal area and lose ability to move purposefully to exit. One of the means that increases the possibility of timely evacuation is the crowd management performed by specially trained intellectual agent-rescuers who can identify high-density areas, reach them in the shortest paths, taking into account obstacles and movements of other rescuers, and eventually get people out of the danger zone.

Often, combined models are used that implement several methods at once. For example, the article Abrosimov and Lebidko (2013) considers the problem of operational management of significant number of people by a limited group of service personal members during mass events. The processes of buying tickets, passing people through control points and other management activities are modeled as the functioning of a queuing system. To describe human behavior, the agent-based modeling is used, but the term "intellectual agent" covers in this case not only a specially trained service personal member, but also a visitor to a mass event. The latter has the ability to assess emerging situations, interact with other visitors, make independent decisions and move in space in accordance with own goals and preferences, depending on the current situation, taking into account the information available in his or her knowledge base or received from other agents.

Simulation of management of an organized crowd, whose members (agents) are united by one object of attention and can act or be passive, is considered in the book Breer et al. (2016). Behavior of the crowd is described as a collective decision-making; for this goal, both the activity of agents and communications between them are significant. These factors are considered within an active network structure. Static and dynamic, deterministic and stochastic management models are used. Results of the simulation are applicable to substantiate measures for "physical" influence on the crowd in order to prevent stampedes, mass riots, etc. The book is also valuable for its extensive bibliography.

2.4 Vehicle Occupants in Crashes

2.4.1 Features of Crash Processes

An important feature of crashes of airplanes, helicopters, cars, and trains is inability of human operator (a pilot or a driver) to control the physical processes with high speed and short duration, which accompany the impact and destruction of vehicles, as already noted in Sect. 2.1. The duration of the first (main) impact as shown by experimental data and simulation results (some are given below) is a fraction of second, and it is less than the delay time of human reaction (see Sect. 2.2.1). Therefore, occupants of the vehicle (crewmembers and passengers) during the impact are subject to almost exclusively the laws of mechanics—similar to elements of the structure, components of propulsion system or equipment and transported cargo.

Given this feature, we will not consider shipwrecks: they develop over a longer period of time, when control of the situation can be maintained—at least partially to ensure the rescue of people. To account for and control human–machine interaction in such situations, the approaches presented in Sects. 2.2.3 and 2.3 are applicable.

Mechanical loads acting on the vehicle structure, equipment and cargo in a crash exceed, as a rule, the design limit values, so large plastic deformations in their elements arisen, significant changes in shape, and other types of loss of load bearing capacity occur. Collapse (destruction) of the vehicle—some elements or the entire structure—is another characteristic feature (along with the impossibility of human operator control) of the situation under consideration.

The duration of the main impact is much shorter than the entire time interval within which the emergency situation connected with vehicle crash develops, but it is this phase, despite its short duration, that covers the destruction processes and determines their possible consequences in the form of fire, explosion, flooding of the vehicle, death or injury of people. This phase should be considered one of the most important for ensuring safety.

The limits of human tolerance to mechanical impacts differ from the limits accepted in the design of vehicle elements. Because of the possible dangerous consequences of destruction, it is necessary to evacuate people from the vehicle immediately after the crash (at the end of the main impact or a series of subsequent secondary impacts). To ensure evacuation, the relative integrity of the cockpit or cabin structure and the functioning of exits are required. In addition, special rescue equipment (inflatable ladders, life jackets, rafts, stretchers) shall be provided for, the means of informing rescuers about the crash site, etc. Thus, taking into account human factors, specific survival criteria arise that go beyond the safety requirements in terms of structural strength. Survival management, aimed at meeting these criteria, should be implemented in the vehicle design stage.

2.4.2 Ensuring Survival in the US Military Aviation

The concept of a design that shall ensure survival in aircraft crashes—*Crashworthiness* (i.e., the design of an aircraft system according to the criteria of safe destructibility in such situations)—has appeared in the USA in the 1970s in connection with the creation of military helicopters UH-60 *Black Hawk* (Sikorsky) and AH-64 *Apache* (Boeing). In order to implement crashworthiness properties in the event of aircraft collision with the ground, data from relevant accidents and catastrophes have been investigated, experiments and calculations using mathematical models have been carried out. The studied survival conditions included accelerations during impact, deformations of the life volume inside the aircraft, the probability of injury to human, etc. Risk of injury was being assessed based on mechanical parameters (forces, accelerations, deformations).

In order to summarize the results of R&D, U.S. Army has organized preparation of the Aircraft Crash Survival Design Guide (ACSDG) in five volumes:

(1) Volume I—Design Criteria and Checklists (USAAVSCOM TR 89-D-22A 1989);
(2) Volume II—Aircraft Design Crash Impact Conditions and Human Tolerance (USAAVSCOM TR 89-D-22B 1989);
(3) Volume III—Aircraft Structural Crash Resistance (USAAVSCOM TR 89-D-22C 1989);
(4) Volume IV—Aircraft Seats, Restraints, Litters, and Cockpit/Cabin Delethalization (USAAVSCOM TR 89-D-22D 1989);
(5) Volume V—Aircraft Post-crash Survival (USAAVSCOM TR 89-D-22E 1989).

In addition, the military standard (MIL-STD-1290A 1988) has been issued. It contains the minimum requirements for ensuring the survival of occupants of light fixed and rotary-wing aircraft in crashes, which must be met in the design stage. Updated design criteria for military helicopters that correspond to modern ideas developed in accordance with the concept of crashworthiness are given in the report Bolukbasi et al. (2011).

In the USA, in addition to the Ministry of Defense, FAA and NASA are engaged in R&D and rulemaking in the field of aircraft safety. An overview of their approaches to create airplanes and helicopters capable of ensuring the survival of crewmembers and passengers in crashes is given in Sect. 2.4.4. Before that, we will consider the work carried out in this field in Russian (former Soviet) aerospace industry in the 1980–1990s.

2.4.3 The Soviet and Russian Research and Development

In the Soviet Union and later in Russia, the work known to the author on ensuring safety in the case of aircraft crash has been performed within the frame of the *Energia-Buran* transport space system project. As one of the postulated abnormal situations for aircraft or spacecraft of aviation type (the *Buran* spacecraft belonged to this type, see Fig. 2.9), relevant regulatory documents envisage emergency landing with landing gear retracted, called by Russian pilots “landing on the fuselage”.

During such a landing, destruction of attachment units for items of cargo (in the terminology of space engineering “payload”) placed in the cargo compartment (payload bay) is possible, impact of these items (masses) on the crew compartment wall, and movement of equipment components torn from the attachment points within the crew compartment. In order to ensure safety of the crew of aircraft or spacecraft, regulatory documents specify the required strength of attachment systems for payload or equipment. In the design stage of *Buran*, the regulatory requirements were used which coincided in regard to emergency landing with provisions of the airworthiness standards for transport category airplanes (ENLG 1985). According to the established requirements, “it should be assumed that design loads applied at the center of gravity of each mass correspond to the following range of acceleration, in g units:

- for longitudinal load—from zero to 9 in the forward direction of the load and from zero to 1.5 in the backward direction of the load;
- for normal (vertical) load—from zero to 4 in the downward direction of the load and from zero to 2 in the upward direction of the load;
- for side load—from +2.25 to −2.25”.

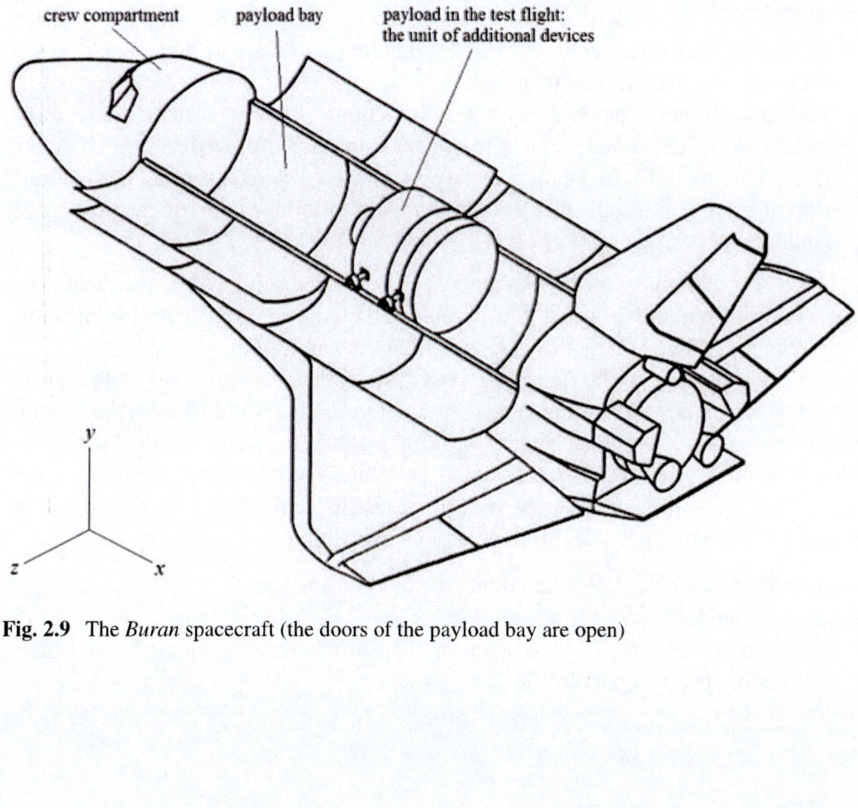

Fig. 2.9 The *Buran* spacecraft (the doors of the payload bay are open)

Taking into account the possible emergency landing on water, it is required to "ensure local strength of the fuselage and those hatches, windows and doors, destruction of which does not ensure the buoyancy of the aircraft during the time required for the passengers and crew to leave the aircraft" (ENLG 1985).

Engineers of NPO *Energia*, participated in development of the *Energia-Buran* system under the leadership of Dr. Viktor F. Gladkii, the head of the department on dynamic loads, have identified the following shortcomings of the regulatory approach to safety:

1. Only a part of the factors threatening the crew during emergency landing is under regulation. Many hazards remain outside consideration. The such are the impact accelerations acting on crewmembers, deformations of the life volume of the crew compartment, the possibility of collapse of the payload structure (while maintaining the integrity of attachment units)—because the payload may contain flammable, toxic or radioactive substances, the possibility of fire and explosion of the spacecraft due to damage of the propulsion system or fuel tanks.
2. The regulatory-specified design loads, based on acceleration values obtained from experimental data on emergency landings of various airplanes and helicopters, are not reliable enough in relation to an aerospace vehicle of an unconventional configuration, such as *Buran*. These values do not take into account dependence on features of the vehicle structure (large transverse dimensions of fuselage, specific design of payload bay, etc.), attachment scheme of the payload, flexibility of the attachment units and dynamic characteristics of the payload structure, mechanical properties of the landing surface and parameters of the vehicle movement at the impact.
3. The airworthiness standards do not define ability of crew or automatic control system of the aerospace vehicle to control conditions for performing an emergency landing until the landing surface is touched. There is no data on relationship of these conditions with hazard factors to minimize consequences of crash landing.

These shortcomings can be explained by the absence in TsAGI, the creator of airworthiness standards, during the period of their development, of methods for simulating a possible emergency (crash) landing when designing an aircraft.

In the design stage of the *Energia-Buran* system, the necessary methodology has been created[20]; it has formed the main content of the author's PhD thesis (Spirochkin 1987). Simulation of the crash landing of *Buran* was based on mathematical modeling of the dynamic system "aerospace vehicle structure + equipment + payload + the landing surface at the contact zone with the fuselage" and solving a derived system of differential dynamic equilibrium equations describing:

- movement of the vehicle when it hits the landing surface;
- structural vibrations of the whole system;
- large plastic deformations, changes in shapes of structural elements and their destruction under impact loads;

[20] In 1983–1987, in NPO *Energia*.

- variable conditions of the contact.

The mathematical model was being built using the finite element method (FEM) in physically and geometrically nonlinear statement. The dynamic equilibrium equations, expressed in incremental form, were being solved by numerical time integration based on implicit schemes. Simulation of the crash landing was being performed with various landing surface types, including regular concrete runway and soil. Finite elements of a special type were imitating the contact of the fuselage with the landing surface. For the case of landing onto soil, their properties have been assigned on the basis of experimental data accumulated during the test landings of the Soviet reentry capsules *Vostok*, *Voskhod* and *Soyuz*.

In order to obtain features of the sliding high-speed impact of the *Buran* fuselage bottom on the landing surface, an experimental facility has been proposed (Spirochkin 1988), then patented (Patrushev et al. 1989).

The finite element model of the analyzed dynamic system has had about 1800 degrees of freedom. Dynamic characteristics of this model (natural vibration frequencies and mode shapes) have been verified by results of the horizontal frequency tests of the *Buran* flying prototype, in the payload bay of which the unit of additional devices has been placed (see Fig. 2.9). The described FEM modeling, calculation of the dynamic characteristics and simulation of the structural behavior of the dynamic system during the crash landing were being carried out using a software package developed by the author and his colleagues; later it became known as *NewTone* (Spirochkin 1993).

As result of this work, the shortcomings of the regulatory approach in relation to *Buran*, indicated above, were eliminated. With the use of mathematical modeling and simulation, it became possible to

- consider the whole spectrum of hazardous factors for an aerospace vehicle during emergency landing in the design stage;
- determine loads acting on the attached masses (payload items, equipment components, crewmember seats) taking into account their placement, attachment scheme, structural features and mechanical properties, as well as the interaction with the vehicle structure;
- evaluate acceptability of these loads (according to the airworthiness standards, the main hazard factors) depending on the emergency landing conditions, which can be controlled in the time interval preceding the landing impact, if the emergency situation is recognized by the pilots and they are able to perform certain actions to soften the impact;
- find the emergency landing conditions that correspond to the levels allowed by the airworthiness standards.

One of these conditions and, accordingly, the actions of pilots is the implementation of emergency landing on the ground (soil) instead of a concrete runway. Analysis of relationship between the loads during the main impact and mechanical properties of the landing surface resulted in recommendations for special preparation of a soil runway, which ensure that the loads acting on the attached masses during emergency

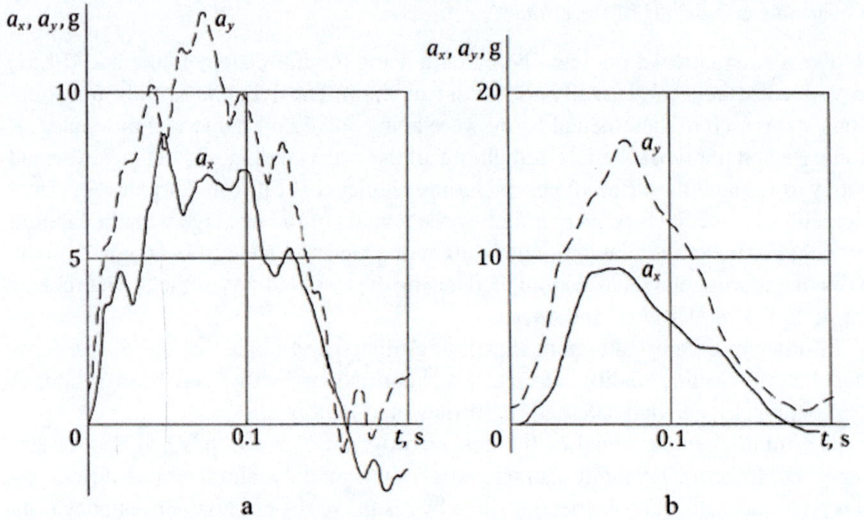

Fig. 2.10 Longitudinal (a_x) and vertical (a_y) accelerations versus time during the emergency landing of *Buran* on the soil runway with ultimate strength 1.1 MPa, with vertical velocity 3 m/s: **a** at the fuselage longeron, **b** at the center of gravity of the payload

landing would not exceed the allowable values. These recommendations included reducing the load bearing capacity and the coefficient of friction of the landing surface to certain limits by plowing and wetting the soil. The spare soil runways with such parameters were to be prepared at the main landing sites for *Buran*: three airfields of the first class on the territory of the USSR.

As simulation of the *Buran* emergency landing onto "soft" soil has shown, the loading process of the payload bay is characterized by the pulses of longitudinal and vertical accelerations with duration about 0.16 s determined by the lower modes of structural vibration of the fuselage combined with vibrations of the load bearing elements with frequencies 50–100 Hz (see Fig. 2.10a). Dynamic loading of the payload itself (the unit of additional devices) is caused mainly by oscillations as a rigid body relative to the attachment points (Fig. 2.10b).

However, due to the significant instability of mechanical properties of the soil, their dependence on climatic conditions and weather, as well as other factors, the possibility to control loads by performing emergency landing on a soil runway is limited. More reliable is reduction of loads by controlling the spacecraft movement parameters, mainly vertical landing velocity and the angle of attack.[21] Dependences of accelerations on these parameters during the *Buran* emergency landing were being determined by relevant simulations. Their results shown that accelerations do not exceed the by-standards-allowed values (see above) if emergency landing is performed:

[21] Because of low values of vertical landing velocity of *Buran*, the angle of attack would not significantly differ from the pitch angle, directly controlled by the pilots.

- on the low-strength soil runway (with ultimate strength no more than 0.4 MPa) with vertical velocity (at the moment of touch) approximately 1 m/s,
- on the concrete runway with vertical velocity about 0.7 m/s.

The angle of attack at emergency landing should be close to the allowed minimum ($\alpha = 9°$).

On the basis of the calculated shock response (acceleration) spectra, requirements for the stiffness of attachment units for low-mass equipment and payload components have been determined. The required stiffness should be such that the frequencies of their partial oscillations (relative to the attachment points) are at least 35 Hz. A study has been carried out, how the dynamic characteristics of heavy masses (equipment and payload) affect the loading of their attachment units during emergency landing. As a result, the possibility of gradation of these masses depending on the value of the mass, also taking into account the stiffness of their structure, has been revealed.

Based on the simulation outputs, the required load bearing capacity levels of the attachment units of the *Buran* power supply system have been updated. The components of this system located in the zone of probable destruction of the spacecraft structure could be collapsed. The update has made it possible to reduce the probability of fire and explosion during emergency landing.

The developed methodology for analyzing the dynamics of the structure, taking into account destruction of the load bearing elements, was used later to investigate the emergency landing, actually crash, of the Mi-34 helicopter by order of TsAGI (Lyakhovenko et al. 1989; Spirochkin and Shenk 1990). It was applied also to study nonlinear dynamic processes in the structure of the International Thermonuclear Experimental Reactor, ITER (Spirochkin 1993). As for the helicopter, the following structural response parameters, characterizing its crash landing, have been obtained:

- the maximum vertical acceleration at the cabin floor is about 40 g (the pulse duration is approximately 0.075 s);
- the maximum nearing of the cabin ceiling and floor in the area of pilots' seats is 12 cm, the residual one is about 3 cm.

These data are consistent with the requirements on survival conditions that must be met in the design of a helicopter; in addition, they correspond to the results of experimental studies. Thus, the standard (MIL-STD-1290A 1988) specifies the maximum vertical acceleration 46 g for a helicopter hitting the ground with velocity up to 15 m/s. In a full-scale physical simulation of such impact with velocity 7 m/s, for 4-ton-helicopter, acceleration 45 g has been registered at the cabin floor, the maximum deformation of the cabin was 15 cm, and the residual one was 9 cm (Wittlin 1973).

Soviet and Russian R&D aimed at ensuring safety in aircraft crashes gradually came to naught. The *Energia-Buran* program was discontinued in the early 1990s. The collapse of the USSR and subsequent socio-economic reforms in Russia proved fatal for the activities in this area: in the absence of funding, the work was being continued for some time only by a small group of specialists under the leadership of Dr. Iosiph A. Lyakhovenko at TsAGI.

2.4.4 FAA and NASA Approaches to Aircraft Crashworthiness

Safety of passengers and aircraft crews in possible aircraft crashes is one of the goals of activities of FAA as the regulatory organization of the USA in the field of civil aviation. The requirements for the design of transport category airplanes taking into account crash landing (or emergency landing) are given in the *Airworthiness Standards* (CFR Part 25, §§ 25.561–25.563, Emergency landing conditions). Relevant provisions are reproduced in Russian *Aviation Rules* (AP-25 2009) and in European Airworthiness Code (CS-25 2018). In accordance with (CFR Part 25, § 25.561):

"(a) The airplane, although it may be damaged in emergency landing conditions on land or water, must be designed as prescribed in this section to protect each occupant under those conditions.

(b) The structure must be designed to give each occupant every reasonable chance of escaping serious injury in a minor crash landing when:

    (1) Proper use is made of seats, belts, and all other safety design provisions;
    (2) The wheels are retracted (where applicable); and
    (3) The occupant experiences the following ultimate inertia forces acting separately relative to the surrounding structure:
        (i) Upward, 3.0 g
        (ii) Forward, 9.0 g
        (iii) Sideward, 3.0 g on the airframe; and 4.0 g on the seats and their attachments.
        (iv) Downward, 6.0 g
        (v) Rearward, 1.5 g.

(c) For equipment, cargo in the passenger compartments and any other large masses, the following apply:

    (1) Except as provided in paragraph (c)(2) of this section, these items must be positioned so that if they break loose they will be unlikely to:
        (i) Cause direct injury to occupants;
        (ii) Penetrate fuel tanks or lines or cause fire or explosion hazard by damage to adjacent systems; or
        (iii) Nullify any of the escape facilities provided for use after an emergency landing.
    (2) When such positioning is not practical (e.g., fuselage mounted engines or auxiliary power units), each such item of mass shall be restrained under all loads up to those specified in paragraph (b)(3) of this section. The local attachments for these items should be designed to withstand 1.33 times the specified loads if these items are subject to severe wear and tear through frequent removal (e.g., quick-change interior items).

(d) Seats and items of mass (and their supporting structure) must not deform under any loads up to those specified in paragraph (b)(3) of this section in any manner that would impede subsequent rapid evacuation of occupants".

As per (CFR Part 25, § 25.562), seats and seat belts must successfully complete dynamic tests on decelerations corresponding to emergency landing and measured on the floor of the passenger compartment or crew cabin. Alternatively, their suitability must be demonstrated by "rational analysis based on dynamic tests of similar type seat" for the same emergency landing conditions. "The seat must remain attached at all points of attachment, although the structure may have yielded … to the extent they would impede rapid evacuation of the airplane occupants".

Each passenger or crewmember "must be protected from serious head injury"—"where head contact with seats or other structure can occur". Such injury is assessed using the Head Injury Criterion (HIC):

$$\mathrm{HIC} = \left\{ (t_2 - t_1) \left[\frac{1}{(t_2 - t_1)} \int_{t_1}^{t_2} a(t)\mathrm{d}t \right]^{2.5} \right\}_{\max} \tag{2.1}$$

where

t_1 is the time moment corresponding to the beginning of the head strike;
t_2 is the time moment corresponding to the strike end;
$a(t)$ is the total acceleration versus time curve for the head strike (t is in seconds, and a is in the units of free fall acceleration g).

The value of this criterion should not exceed 1000. In addition, quantitative limitations for loads on the lumbar column and the femurs are established if they can be injured by contact with seats or other structural elements.

§ 25.563 of (CFR Part 25) relates to the case of emergency landing with immersion in water (ditching). This section refers to other sections of the *Airworthiness Standards* that regulate measures to minimize the probability of injury and eliminate obstacles for evacuation of people, as well as the use of life-saving equipment, etc.

Thus, the FAA's regulatory approach takes into account the possibility of destruction of aircraft elements during emergency landing and determines the need to protect people on board. However, ensuring the survival through the safe destructibility of the structure is not regulated directly. This is due to the dependence of the design requirements, including those expressed in the quantitative criteria above, on the design parameters of aircraft. The established requirements are based on an implicit assumption about the typical nature of airframe structure of transport airplanes, which is "constructed predominantly of metal, using skin-stringer-frame architecture" (Federal Register 2015). The used strength criteria correspond to the experience gained for conventional aircraft configurations. They do not cover behavior of non-metallic materials or non-standard structural schemes. In the case of licensing an airplane with unconventional fuselage structure (on materials and/or design), FAA prescribes special conditions for providing a crash protection, the level of which must be "equivalent to that provided by a traditionally-configured metallic airplane"

(Federal Register 2015). These special conditions are comparative in nature, and with respect to safe destructibility, they cannot be considered invariant to design performance.

To address this issue and develop more general requirements for crashworthiness, FAA carries out research work covering a wide range of aircraft, from commercial transport airplanes to general aviation, including helicopters. When carrying out R&D, FAA cooperates with the U.S. Army and Navy, various government organizations and universities. An important partner is the NASA Langley Research Center in Hampton, Virginia.

At the end of the twentieth century, the Aircraft Crashworthiness Research Program (ACRP) has been launched (Frings 1998; FAA 2001). Its goal was to reduce injuries and prevent deaths of people inside aircraft in severe but survivable crash impact conditions through a systematic approach to crashworthiness. The systematic approach provided for:

- improving aircraft structure in terms of crashworthiness;
- creating a database that would accumulate test and accident information on the crash characteristics of aircraft structures, cabin/cockpit interiors, and occupant seat/restraint system;
- developing analytical methodologies to predict aircraft response and occupant survivability under generalized crash impact conditions;
- ensuring that the R&D results will meet FAA's requirements with minimal costs.

Research was being carried out in the following areas (Frings 1998):

(1) cabin interiors;
(2) fuel systems;
(3) crash environment (crash impact conditions);
(4) aircraft structures;
(5) occupant protection;
(6) analytical/numerical modeling and simulation.

Work in the first of these areas was aimed at minimizing (ideally eliminating) injuries and fatalities caused by cabin equipment (overhead stowage bins, seats, galleys, etc.), based on quantitative assessments of the equipment response. To obtain response data, vertical drop tests with airplane fuselage sections were being carried out on a special facility at the FAA William J. Hughes Technical Center, at the Atlantic City International Airport, New Jersey. Figure 2.11 shows a general view of this facility and one of the test objects—a Boeing 737 fuselage section with a length of about 3 m. It has 18 seats occupied by dummy passengers, the overhead bins with a part of the luggage, and the other part was stowed in the luggage compartment under the cabin floor. The vertical velocity at the impact on the ground was approximately 10 m/s. The response parameters were being recorded using strain gages, accelerometers and video cameras (Byar et al. 2001).

Such work became one of the components of international activities on the Cabin Safety Research Program, which involved, along with the FAA, the Directorate of

a b

Fig. 2.11 **a** The FAA's facility for vertical drop tests (Reproduced from Abramowitz et al. (2000)), **b** General view of the Boeing 737 fuselage section before the test (Reproduced from Jackson and Fasanella (2001))

Transport Canada Aviation and European association of aviation regulators—the Joint Aviation Authorities (Koenig 1996).

Fuel systems (the second area of ACRP) were being investigated in order to reduce risk of post-crash fires caused by destruction of the aircraft components that are crucial for survival. This area covered transport airplanes and helicopters. For transport airplanes, the following work was being performed (Frings 1998):

- vertical and horizontal impact tests of airplane fuselage sections equipped with auxiliary fuel tanks;
- looking for crash resistant fuel system solutions;
- development and testing of fuel containment concepts.

Based on the results of this work carried out for helicopters, the Advisory Circular 29-2B used at certification of rotorcraft has been updated in provisions related to the fuel systems resistant to destruction in impacts (Frings 1998). The current version of this document is (AC 29-2C 2018).

When studying crash impact conditions (the third area of the ACRP), survival limits were ascertained depending on aircraft types, including transport airplanes, helicopters and small airplanes. Generic characteristics for all these types were also considered.

For transport airplanes, the research comprised (Frings 1998):

- ditching conditions;

- evaluating current standards on dynamic tests;
- identifying applicability of the existing design standards to new aircraft types and the need for revision;
- the use of digital cameras to record crash impact parameters.

For helicopters, the research was performed as a joint project with the US Navy: it covered cases of ditching. Conditions of rotorcraft rollover were studied, and crashworthiness requirements for medevac (intended for medical and evacuation goals) helicopters were determined.

The research regarding the small air vehicles was focusing on air taxi and similar aircraft ditching, the need to review the design standards and the regulation of full-scale impact tests.

To obtain generic aircraft characteristics, dependencies of response parameters in crash impact on aircraft dimensions were evaluated.

The purpose of work on aircraft structures (the fourth area of ACRP) was to determine loads in crash impact, and fuselage response for various air vehicles. Research encompassed small airplanes (in cooperation with NASA), helicopters (in cooperation with the U.S. Army on energy-absorbing structure parts) and unique aircraft structures (Frings 1998). Crashes were being imitated on the FAA's test facility by vertical drop tests of full-scale airplanes (McGuire and Vu 1998), (Abramowitz and Vu 2008), as shown, for example, in Fig. 2.12.

Fig. 2.12 Vertical drop test of the Shorts 330 regional transport airplane. Reproduced from Abramowitz and Vu (2008)

Aircraft landing impact tests also were being carried out on a more powerful facility at NASA Langley Research Center in Hampton, Virginia. This facility, a more than 72-m-high gantry structure, was built in 1963 to practice lunar landings in the frame of the *Apollo* Moon Landing Program. In 1972, the facility was converted to carry out full-scale crash tests with light airplanes and helicopters. Since that time, the facility was being used to perform drop tests for aviation industry, civil aviation, the U.S. Department of Defense, and NASA space programs (Jackson and Fasanella 2003). After a number of improvements for new tasks (in the period from 2004 to 2011), it received its current configuration and name: Landing and Impact Research Facility (LandIR). The features of LandIR are the ability to control simultaneously both the spatial position and velocity of the object tested, as well as to perform crash tests with various combinations of horizontal and vertical velocity conditions due to a pendulum suspension. The tests carried out at this facility and other NASA approaches to aircraft crashworthiness will be described further; in the meantime, we return to the overview of the ACRP program.

Its fifth area covered research and development in the field of protection of passengers and crews of all aircraft types, including (Frings 1998; FAA 2001):

- refinement of head injury criteria (including HIC)—together with Wichita State University (WSU);
- revision of existing and development of additional occupant injury criteria;
- design of energy-absorbing seat cushions—with participation of the FAA's Civil Aeromedical Institute (CAMI).

For helicopters, the use of inflatable airbags in the cockpit was being studied together with the U.S. Army; tests were being conducted on the NASA's LandIR facility (see below). For small airplanes, means for protection of occupants in the side seats were being considered (in cooperation with CAMI).

In research on analytical modeling (the sixth area of ACRP), computational methods and software tools were being developed and tested that could predict damages in aircraft crash, loads acting on the structure and seats, as well as on passengers. The initially used program KRASH (developed by Lockheed-California Company), based on semi-empirical modeling of aircraft structure by concentrated masses, nonlinear springs and beams, was replaced by more advanced analysis tools based on FEM—such as MSC.Dytran (FAA 2001; Fasanella and Jackson 2001).

Description of the work in this area can be found in the paper Jackson and Fasanella (2001). It concerns numerical modeling of the Boeing 737 fuselage section and simulation of its vertical drop test (see Fig. 2.11). Structure of the section was modeled by more than 13,600 shell and beam finite elements; equipment and dummies were imitated by lumped masses concentrated in the corresponding nodes of the finite element model. The simulation was performed with the use of the MSC.Dytran program implementing an explicit method for solving nonlinear dynamics equations. The time integration step was chosen approximately 2×10^{-6} s. The calculated acceleration histories (changes in time) obtained by the simulation were filtered to "clean" from high-frequency noise. Duration of the acceleration pulses at the floor level was 0.1–0.15 s; on the overhead bins it increased up to 0.3 s. Maximum values of the accelerations ranged from 14 to 22 g; their differences from experimental data reached 25%. The calculated acceleration processes were significantly different from those recorded during the test.

As part of an interagency agreement, FAA cooperated with the National Highway Traffic Safety Administration (NHTSA). The cooperation included the exchange of information on the behavior of human body during crash and criteria for injuries. Interaction with universities promoted a systematic approach to crashworthiness. Thus, Drexel University was involved in analytical modeling of the vertical drop test with the Boeing 737 fuselage section (see Fig. 2.11), and the Cranfield Impact Center (CIC) at the University of Cranfield, UK, was participated in supporting similar tests with small airplanes. Interaction with the army provided validation and improvement of analytical numerical models based on experimental data. Common interests of FAA and U.S. Navy laid in the research on helicopter crashes, including impact conditions and crash simulation.

Analytical modeling of the vertical drop test of the Boeing 737 fuselage section was being performed at Drexel University in parallel with the work described above (Jackson and Fasanella 2001). The LS-DYNA program was being used; it, like MSC.Dytran, belongs to the family of explicit nonlinear codes. The finite element model of the tested section has included more than 56,700 shell and beam elements (Byar et al. 2001). Figure 2.13 shows a three-dimensional (3D) geometric model, which has served as an intermediate object for building this finite element model (by detailed meshing).

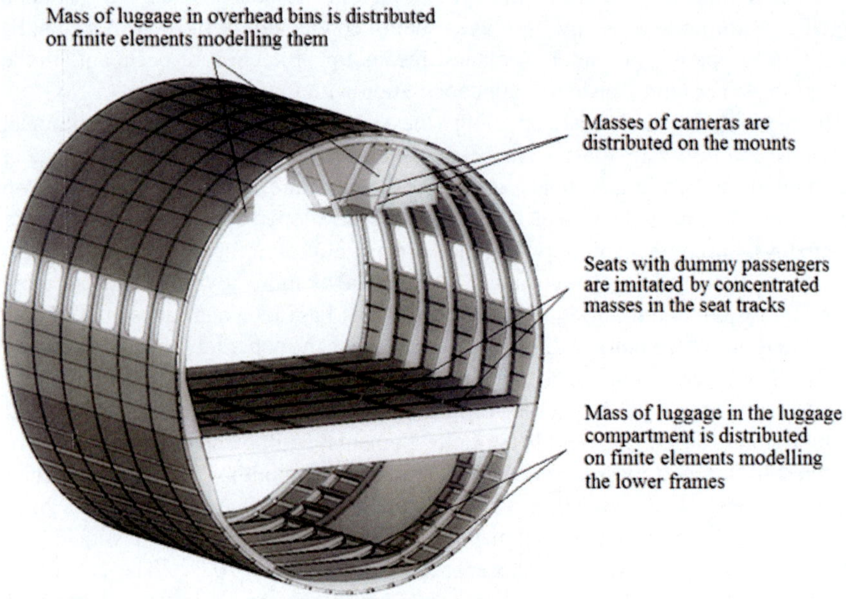

Fig. 2.13 3D geometric model of the Boeing 737 fuselage section, the basis for its finite element modeling. Reconstruction based on data from Byar et al. (2001)

Structural dynamics parameters (displacements, accelerations, internal forces) were being calculated for the time interval 0.35 s with the integration step 10^{-6} s. The calculation results are presented in the reviewed publication, however, in a shorter interval—up to 0.1 s or up to 0.2 s, and not all parameter histories have their maximums, which makes it not possible to assess the amplitude and duration of impact pulse. Despite the very small integration step, only 2–3 sinusoidal vibration components with frequencies approximately from 25 to 50 Hz appear in the dynamic response. The calculated histories of accelerations differ from the results obtained in the crash simulation of the same test using MSC.Dytran (see above). It is obvious that the question of adequacy of finite element models and software tools used in both cases requires additional attention.

Regulatory framing of FAA's requirements to aircraft crashworthiness design is carried out taking into account recommendations of the Aviation Rulemaking Advisory Committee (ARAC). In 2015, ARAC was assigned to prepare recommendations for development of standards for aircraft crashworthiness during emergency landings on land and water and incorporate them into CFR Part 25 (Federal Register 2015). To date, these standards are not yet been published.

NASA's approaches to aircraft crashworthiness in emergency landing include:

(1) full-scale drop tests of almost complete small airplanes, airplane sections and helicopters on the above-mentioned LandIR facility;
(2) controlled crash landing of a real transport airplane;
(3) development and testing of computer software for crash landing simulation.

The purpose of full-scale drop tests is to obtain experimental data characterizing aircraft structural dynamics, and loads transmitted to passengers and crewmembers during crash landing. These data could be used for verifying analytical calculation results, as well as for evaluating the design concepts of aircraft structure, seats and restraints ensuring survival under conditions of destruction. From 1974 to 2003, the following tests were conducted: 41 tests of small civil airplanes, 59 full-scale tests of helicopters (11 for qualification of helicopters taking into account crash, and 48 for qualification of Wire Strike Protection System in Army helicopters), 3 vertical drop tests of Boeing 707 fuselage sections, 8 vertical drop tests of a crashworthy composite fuselage section, and more than 50 drop tests of the F-111 crew escape capsule (Jackson and Fasanella 2003).

Based on the test data obtained, impact loading criteria were established, which are used by FAA as standard ones when certifying occupant seats. These data were also included in the Small Airplane Crashworthiness Design Guide (Hurley and Vandenburg 2002).

In drop tests with the AH-1S military helicopter, the active crew restraint systems were being evaluated. These systems included inflatable air bags integrated into a standard five-point restraint harness. The bags were being filled with gas from a gas generator at a signal from a sensor under crash conditions, which could be estimated as severe but potentially survivable. It has been demonstrated that such safety systems can reduce the number of injuries and fatalities caused by impacts on interior cockpit structures. The result of the research was the corresponding modernization of several types of helicopters (Jackson and Fasanella 2003).

If we consider the extended class of active safety systems, they can include the F-111 crew escape capsule, presented in Sect. 2.2.3. Landing impacts of such capsules on the ground (in real combat conditions and during tests) led to a high level of injuries due to collisions of pilots with details of cabin interior. Consequently, an airbag-based impact energy absorption system was developed. It was mounted on the flat bottom of the capsule outside (see Fig. 2.14) and was equipped with a valve regulating amount of absorbed energy (Jackson and Fasanella 2003; Jackson et al. 2006).

Drop tests of composite fuselage sections were and continue to be part of NASA's exploratory research on safety in aircraft crashes. The objects of such research include concepts of energy absorbing by (Jackson and Fasanella 2003):

- cabin or cockpit subfloor which has a special structure made of composite materials;
- fuselage frame, in which load bearing elements made of composites can be deformed plastically, like traditional aluminum parts;
- seat cushions—the last barrier protecting aircraft occupants in accident, with a variety of foam materials.

One of the recent experiments at LandIR was a crash test of a full-size Fokker F-28 transport airplane carried out in the summer 2019 in partnership with FAA

Fig. 2.14 Preparation for a test of the F-111 crew escape capsule with airbag-based safety system. Reproduced from Jackson et al. (2006)

(NASA 2020). The 15-ton airplane (the largest ever tested on this facility), equipped with 24 human dummies and cameras, was suspended at a height of more than 45 m, and then dropped to the ground (see Fig. 2.15).

The airplane was covered outside with a pattern of dots to provide photogrammetry: by their displacements, deformations of the structure can be tracked. After the crash, spatial scanning was carried out by specialists of the National Transportation Safety Board (NTSB). To get necessary data on dynamic response of passengers and crew when airplane impacts on the ground, CAMI and the U.S. Army Test and Evaluation Command provided various types of anthropomorphic dummies in a range of parameters—from a 3-year-old child to an obese man weighing more than 120 kg (see Fig. 2.16). One of the dummies was a unique device created by the U.S. Army to assess injuries of military personnel, including that in the event of an explosion under a vehicle: Warrior Injury Assessment Manikin (WIAMan).

Another approach of NASA to aircraft crashworthiness (controlled crash landing of a real transport airplane) is presented in the Controlled Impact Demonstration (CID). This joint experiment with FAA was performed in late 1984: a remotely piloted Boeing 720 took off from Edwards Air Force Base, and after a short flight was commanded to come into contact with the ground. The objectives of CID were (McComb et al. 1987; Jackson and Fasanella 2003; NASA Wiki 2016a):

- to demonstrate a reduction of post-crash fire, when fuel tanks are destroyed, through the use of antimisting kerosene (containing a special additive);
- to study behavior of transport airplane structure in crash impact conditions;
- to evaluate the effectiveness of improved seats, seat-restrains and some cabin elements.

In order to record structural reaction parameters, NASA specialists have developed special equipment including 350 sensors that can work even in fire conditions.

The airplane contacted with a specially prepared runway on the east side of Rogers Dry Lake, with the landing gear retracted. Velocity at the impact was: in the horizontal direction about 75 m/s, in the vertical one—more than 5 m/s, there was a roll to the left (see Fig. 2.17a). The wings were sliced open by posts cemented into the runway, and these posts also damaged the engines. All this provided a powerful source of ignition. Data were obtained, despite on the resulting fire (Fig. 2.17b), from 97% of sensors. They were the first quantitative measurements of the structural reaction, performed in the process of a real crash in free flight.

The impact demonstrated that the tested antimisting additive was not sufficient to prevent a large-scale post-crash fire, though it reduced intensity of the initial fire. Researchers from FAA estimated that between 23 and 25% of the 113 people in the plane could have survived the crash. The evaluated time of evacuation of passengers through the front and rear exits was 15 s and 33 s, respectively. This time is basically enough to get to the doors and open them, but the possibility of moving quickly through the thick smoke is problematic. As a result of the analysis, FAA has established new requirements for the flammability of seat cushions and for the mechanical attachment of emergency floor lighting. In addition, the provisions of

Fig. 2.15 Crash test of a Fokker F-28 at NASA's LandIR facility: **a** suspended airplane, **b** falling to the ground, **c** after the crash. Reproduced from https://www.nasa.gov/langley/fokker-f28-crash-test-nasa

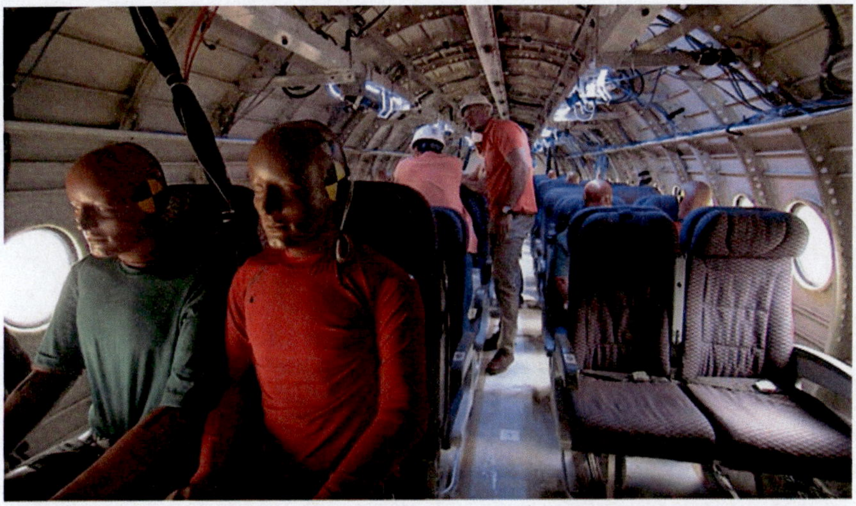

Fig. 2.16 Anthropomorphic dummies inside Fokker F-28. Reproduced from (NASA 2020)

the Federal aviation regulations for flight data recorders regarding the pitch, roll, and acceleration sampling rates have been estimated insufficient (NASA Wiki 2016a).

Crash tests carried out by NASA are usually accompanied by computer simulation of emergency landing impacts. One of the tools for such simulation, used in the period before 1990, was a software program code DYCAST (Dynamic Crash Analysis of Structures). It has been developed at the Grumman Aerospace Corporation with support of NASA and FAA. This code was based on FEM and implemented calculation of structural dynamics parameters taking into account plastic flow and large deformations. For the calculation, unlike the above-mentioned programs MSC.Dytran or LS-DYNA, implicit integration methods were used. It was through the DYCAST code that computer simulations of the CID experiment were carried out. The finite element model was built from 126 beams, 73 springs approximating to be destroyed parts of the structure, 15 springs representing contact of the fuselage with the ground, and 113 concentrated masses. It included 196 independent and 68 dependent degrees of freedom. The time step was 0.0005 s, and the impact duration according to the simulation was approximately 0.15 s (Fasanella et al. 1990). Figure 2.18 shows a comparison of the vertical accelerations on the cabin floor obtained by the simulation with the corresponding experimental data. In general, the simulation of the CID experiment showed a good correlation with the actual data (Jackson and Fasanella 2003).

The analysts at the NASA's LandIR facility used besides DYCAST other software, for example, DYNA3D and NIKE3D codes developed at the Lawrence Livermore National Laboratory with sponsorship of the U.S. Department of Energy. These codes calculated high-speed impact processes using explicit (DYNA3D) and implicit (NIKE3D) methods. As a result of modifications of the open version of the first code

Fig. 2.17 Controlled ground impact of the Boeing 720: **a** initial ground touch, **b** fire. Reproduced from (NASA Wiki 2016a).

Fig. 2.18 Comparison of the vertical floor accelerations on the Boeing 720 cabin floor obtained experimentally and by simulation. Reproduced from Fasanella et al. (1990)

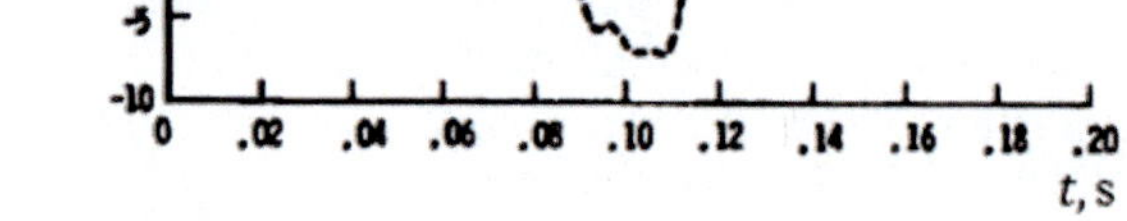

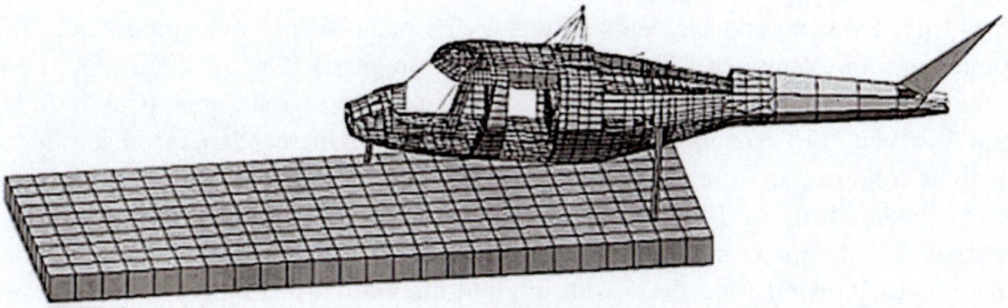

Fig. 2.19 Finite element model of a Sikorsky ACAP helicopter in contact with the ground intended for MSC.Dytran. Reproduced from Jackson et al. (2006)

(made by its vendors), commercial programs MSC.Dytran and LS-DYNA as well as PAM-CRASH appeared.

The MSC.Dytran program was subsequently used for computational support of drop tests of a prospective Sikorsky ACAP composite helicopter, composite fuselage sections with anthropomorphic dummies, and other similar experiments (Jackson and Fasanella 2003). The finite element models were being verified by results of experiments and were becoming more and more complex (see, e.g., Fig. 2.19). Comparison of the calculations based on such models with experimental data showed their consistency, which in turn led to confidence in explicit nonlinear codes as "a crashworthy design and certification tool for aircraft" (Jackson et al. 2006).

Additional information about FAA and NASA research, including that on energy-absorbing fuselage structural elements and computer simulation tools for analyzing aircraft crashworthiness, can be found in a number of publications, for example: Jackson and Fasanella (1999), Lyle et al. (2000), Jackson and Fasanella (2002), Stockwell (2002), Fasanella et al. (2003), Fuchs and Jackson (2008), Fasanella et al. (2008), Ren and Xiang (2014) and Guida et al. (2018). The latter also contains extensive bibliography.

2.4.5 Other Safety Studies in Aircraft Crashes

Studies on safety of passengers and crewmembers in aircraft crashes are also being carried out in Western Europe. One of their directions is controlled absorption of energy by the aircraft fuselage elements collapsed during impact. In the publication Johnson et al. (1997), describing the state of such works at the end of the twentieth century, the variants of the structural design of the helicopter floor using energy-absorbing composites, technological aspects of manufacturing the corresponding structural parts and experimentally obtained characteristics of their deformation are presented.

The current state is discussed in the articles Guida et al. (2018) and Riccio et al. (2019). Drop tests of fuselage sections, which reveal the features of impact energy absorption by structural elements of the floor (made mainly of composites), are combined with computer simulations of impact processes. One of the leading test organizations in this direction is the Italian Aerospace Research Center (Centro Italiano Ricerche Aerospaziali, CIRA) in Capua. In order to increase energy absorption, the to be collapsed structural elements are provided with "triggers"— features in the form of chamfers or tines that initiate crushing and guide it on the desired path (Guida et al. 2018). Computer simulations are carried out using LS-DYNA, ABAQUS or other codes implementing FEM with implicit integration methods. Finite element models (similar to the one shown in Fig. 2.19) use the deformation characteristics approximating experimental data.

After completion of the ACRP research program (see Sect. 2.4.4), theoretical and experimental studies to ensure protection of passengers and crewmembers in aircraft crashes continued in a number of research centers in the USA and other countries. For example, at WSU, with the participation of the National Institute for Aviation Research (NIAR), influence of temperature on mechanical, in particular viscoelastic, properties of the seat cushions filled with polyurethane foam was studied (Beheshti and Lankarani 2010). Regulations of FAA require dynamic testing to confirm the suitability of seats at room temperature (CFR Part 25, § 25.562). However, crash of an airplane or a helicopter is possible at low ambient temperatures. In addition, viscoelastic properties of foam may depend on temperature variations during the service life of seats. The corresponding changes in properties can affect dynamic behavior of airbags and restraint systems in emergency impact, and such changes are difficult to reproduce in certification tests. To solve this problem, computer simulation was used: the body of an occupant was modeled by concentrated masses, the seat with seat belts—by elastic bonds, and dynamics equations were being solved using an explicit program code Mathematical Dynamic Model (MADYMO). Simulation of mechanical behavior of the foam-filled seat cushions was based on the experimental data obtained in the temperature range from −40 to +50 °C.

The second controlled crash of a full-scale aircraft after the CID experiment (see Sect. 2.4.4) was a Boeing 727 emergency landing carried out in 2012, in the Sonora Desert, Mexico. This crash test was organized by a row of TV channels from the USA and Great Britain: *Discovery Channel*, *Channel* 4, etc.; its cost was approximately 1 million pounds. The goal was to study destruction of aircraft structure, impact loads on human body, and effect of seat belts, to determine the most dangerous places for passengers and to assess chances of survival in similar accidents in the future (Joseph 2012).

In the cabin of the airplane, three anthropomorphic dummies were placed (see Fig. 2.20):

(1) one fastened with a seat belt and tilted in the "emergency" position (in the brace);
(2) one belted but with "straight back";
(3) and one neither belted nor in the brace.

Initially, the airplane was flying under the control of the pilot who was in it. At altitude about 800 m from the ground, the pilot parachuted, and then the Boeing 727

Fig. 2.20 Dummies in the cabin of the Boeing 727. Artistic reconstruction based on data from Joseph (2012)

was guided via remote-control device arranged in a following Cessna for landing into the bottom of a dried-up lake.

The impact on the ground occurred with a negative pitch angle (see Fig. 2.21a) at a horizontal speed 230 km/h and a vertical speed 7.62 m/s. The landing gear was not retracted, which contradicts to the FAA regulation (CFR Part 25, § 25.561). The horizontal force on the front landing gear from the ground caused destruction of the fuselage forward part (Fig. 2.21b). As a result, the cockpit and the front part of the cabin came off (Fig. 2.21c).

All three engines remained intact and continued to work after the crash until they were flooded with water when firefighters arrived at the scene. The fire was avoided. Experts found that the dummy, wearing a seat belt and sitting in an "emergency" position (marked as 1 in Fig. 2.20) would have survived the impact and not injured—this confirms effectiveness of the relevant regulations for passengers. The second belted dummy but with "straight back" (2) would have suffered serious head injuries. The dummy not wearing a seat belt (3) would have perished.

The experts concluded that 78% of passengers on board would most likely have survived the impact. This number does not include pilots and first class passengers: when the nose of airplane takes the brunt of the impact and the front of the fuselage shears off, they have almost no chance of survival.

This conclusion is consistent with the point of view that the probability of survival in a crash depends on the sitting place occupied by the passenger. According to Anne Evans, a former investigator at the UK's Air Accidents Investigation Branch (AAIB), "it is safer to sit at the back of the aircraft where the flight recorder is. The front is more vulnerable because that often sees higher impact forces" (Joseph 2012).

However, this point of view is not shared by either regulatory authorities, in particular, FAA, or aircraft manufacturers, including Boeing. Still, it can be substantiated

Fig. 2.21 Crash landing of the Boeing 727. Artistic reconstruction

by a study conducted by the *Popular Mechanics* magazine on NTSB data relating to crashes of commercial jets in the USA, since 1971, in which there were both fatalities and survivors (Noland 2007). As a result of this study, relative survival rates were determined for various parts of the passenger cabin. Average value of this survival indicator for each of the three selected parts of the cabin is distributed as follows: 69% in the rear cabin (behind the trailing edge of wing), 56% in the center section and a small area in front of it, and 49% in the first and business class (see Fig. 2.22).

Currently, in light airplanes of European production, the front passenger seats do not meet the FAA regulations (CFR Part 25, § 25.562) due to high value of the Head Injury Criterion, HIC (Lamanna et al. 2019). In addition, there are inconsistencies between the American and European certification rules (CS-23 2003). To meet requirements of FAA, there is a need for design changes. Computer simulations and experiments aimed at improvement of seats and restraint systems form another direction of the research, conducted in Western Europe. At simulations, combined models are used: human body is approximated mainly by a system of mechanical entities with concentrated parameters, similar to that described in the publication Beheshti and Lankarani (2010), and the seat structure is modeled by finite elements.

The article Lamanna et al. (2019) presents an analytical study carried out with a numerical model of a seat, the total number of finite elements in which exceeded 105,000. For the elements that modeled aluminum parts of the structure, a deformation diagram with hardening was used; those that represented polyurethane foam deformed like rubber. Dynamic equilibrium equations were being solved by LS-DYNA. Experimental studies were performed using anthropomorphic dummy Hybrid II according to (CS-23 2003). The results, presented in the article, show significant differences in shapes of the acceleration pulses, as well as in the HIC values obtained by computer simulation and by experiment. In particular, the calculated HIC values exceed the experimental ones by more than 1.7 times. Thus, the question of adequacy of analytical modeling remains relevant.

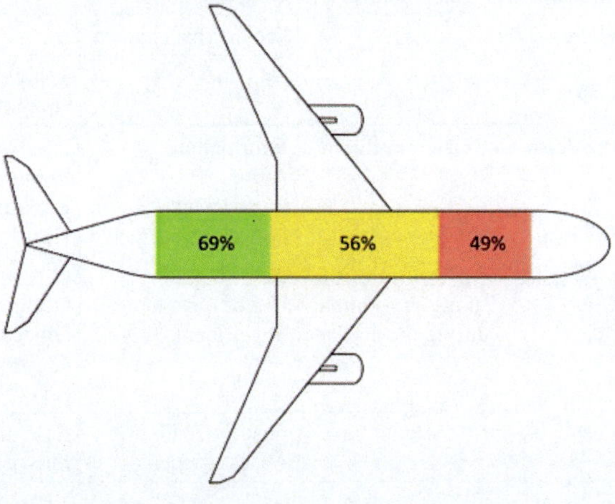

Fig. 2.22 Survival rates for various parts of the passenger cabin. Based on data from Noland (2007)

2.4.6 Car Accidents

A frame approach to ensuring safety of occupants of a car (in general, a ground wheeled vehicle, including bus, truck, etc.) in accidents: collision with an obstacle or other vehicle, falling from great height, and other crashes—was proposed in the 1970s by William Haddon. This approach is described by a matrix of factors that characterize human, vehicle, and environment in various phases of accident (before, during, and after). The Haddon matrix is another example of information models that complement those presented in Sect. 2.2.4. Its current form, adopted by the World Health Organization (Peden et al. 2004), is shown in Table 2.3.

The Haddon matrix can be used to assess relative importance of various factors and measures to reduce fatalities by preventing or minimizing injuries. Design measures for protecting humans during an accident correspond to the matrix cell that is located at the intersection of the row "Crash" and the column "Vehicle and equipment". They are implemented by a special design of the vehicle structure and equipment, including interior, aimed at protecting occupants during the crash impact, as well as by the use of passive and active safety systems. These measures are regulated in all economically developed countries.

In the Russian Federation, the Technical Regulations of the Customs Union "On the safety of wheeled vehicles" (TR TS 018/2011) are in force. Provisions of this document comply with the United Nations Regulations, or the UN Regulations (UN Vehicle Regulations) and the Global Technical Regulations (GTRs 1998) that provide for safety in collisions and other similar road accidents. Requirements of all the above regulations ensure minimization of the traumatic effects on the vehicle occupants during accident and post-crash evacuation; they determine the design of the vehicle and the properties of its elements that affect safety (see Table 2.4).

The UN Regulations contain provisions on the official approval of vehicles concerning the protection of the driver and passengers from dangerous factors in

Table 2.3 The Haddon matrix (Peden et al. 2004)

Phase		Factors that characterize		
		Human	Vehicle and equipment	Environment
Pre-crash	Crash prevention	Information Attitudes Impairment Police enforcement	Roadworthiness Lighting Braking Handling Speed management	Road design and road layout Speed limits Pedestrian facilities
Crash	Injury prevention during the crash	Use of restraints Impairment	Occupant restraints Other safety devices Crash-protective design	Crash-protective roadside objects
Post-crash	Life sustaining	First-aid skill Access to medics	Ease of access Fire risk	Rescue facilities Congestion

Table 2.4 Regulatory requirements on minimizing the traumatic effects on the vehicle occupants and ensuring the possibility of their evacuation after a road accident (TR TS 018/2011)

Vehicle elements and properties	Regulatory documents, compliance with which ensures the requirements to be met
Door latches and door retention components	UN regulation No. 11—Rev. 2, Rev. 3; Global technical regulation No. 1
Protection of the driver against the steering mechanism in the event of impact	UN regulation No. 12—Rev. 4
Safety-belt anchorages	UN regulation No. 14—Rev. 7
Safety-belts, restraint systems	UN regulation No. 16—Rev. 10
Strength of seats, their anchorages and head restraints	UN regulation No. 17—Rev. 6; UN regulation No. 80—Rev. 1
Interior fittings	UN regulation No. 21—Rev. 2
Head restraints (headrests)	UN regulation No. 25—Rev. 1
Protection of the occupants of the cab of a commercial vehicle	UN regulation No. 29—Rev. 2
Safety glazing materials and their installation on vehicles	UN regulation No. 43—Rev. 4
Large passenger vehicles with regard to the strength of their superstructure	UN regulation No. 66—Rev. 1
Protection of the occupants in the event of a frontal collision	UN regulation No. 94—Rev. 3
Protection of the occupants in the event of a lateral collision	UN regulation No. 95—Rev. 2

road accidents, as well as on crash tests for this purpose. When testing (with the exception of tests for the door locks and hinges), physical models of human body parts (head, neck, torso, legs) are used, or dummies, including anthropomorphic test devices that meet the US Federal Motor Vehicle Safety Standards (see below), and the criteria of injury are assessed. Protective properties of the cab and vehicle body structure are checked by the strength of the load bearing parts (racks, roof, etc.), as well as by survival space required after deformation. For large vehicles, the basic method for confirming the strength of the roof in case of rollover is testing, as in relation to other requirements, but there are also equivalent methods: quasi-static calculations and computer simulations.

When testing cars for a frontal collision, values of the Head Performance/Injury Criterion (HPC) are assessed; this criterion is almost equivalent to the HIC (see Sect. 2.4.4), but it is calculated using a slightly different formula, given in UN Regulation No. 94:

$$\mathrm{HPC} = (t_2 - t_1)\left[\frac{1}{(t_2 - t_1)}\int_{t_1}^{t_2} a(t)\mathrm{d}t\right]^{2.5} \tag{2.2}$$

where designations are the same as in expression (2.1). The HPC value shall not exceed 1000 units and the head acceleration shall not be more than 80 g for more than 3 ms. In addition, in UN Regulation No. 94, allowable values are established for

- the Neck Injury Criteria (NIC), including neck tension criterion, neck shear criterion and neck bending moment;
- the Thorax Compression Criterion (ThCC);
- the viscous criterion (VC) for the thorax;
- the femur force criterion (FFC);
- the tibia compression force criterion (TCFC);
- the tibia index (TI) and
- the movement of the sliding knee joints.

The use of Hybrid III dummies that meet American specifications (see below) is prescribed. Vehicle speed at the moment of impact shall be 56 km/h (see UN Regulation No. 94).

To assess injuries in the event of a lateral collision, the criterion HPC is considered; according to UN Regulation No. 95 it is complemented by

- the thorax performance criteria—Rib Deflection Criterion (RDC) and soft tissue criterion VC;
- the pelvis performance criterion—Pubic Symphysis Peak Force (PSPF) and
- the abdomen performance criterion—Abdominal Peak Force (APF).

The values of these criteria obtained from crash tests are compared with the established limits. The speed of the mobile deformable barrier at the moment of impact shall be 50 km/h. The dummy shall correspond with the specification of the ES-2 dummy (see UN Regulation No. 95).

In addition, to minimizing mechanical injuries, in collisions, spontaneous opening or blocking of doors should be excluded and protection against large fuel leaks, electrolyte ingress into the interior, and electric shock should be provided.

Global Technical Regulation No. 1 contains the supplementing requirements for door locks and door retention components, including latches, hinges, strikers, trunk lids, etc., in order to minimize the likelihood of the driver or passengers being thrown out of the vehicle in emergency impact. They are in force along with UN Regulation No. 11 and take into account relevant experiences of the USA, as well as a number of other countries.

For vehicles, there are also requirements for minimizing physical impacts on other road users, including pedestrians. These requirements are not directly related to the people inside a vehicle in emergency impact, but it would be wrong not to pay attention to safety of those who are accident participants outside. The aspects of such situations to be regulated and relevant regulatory documents are presented in Table 2.5.

In the USA, the previously mentioned Federal Motor Vehicle Safety Standards (FMVSS), developed and enforced by the NHTSA, are used. These standards are currently codified in Title 49 of the Code of Federal Regulation (CFR Part 571).

Table 2.5 Regulatory requirements on minimizing the physical impacts on other road users (TP CU 018/2011)

Requirements to be met	Regulatory documents containing the requirements
Reducing the risk or seriousness of bodily injury to a person hit by external projections of the vehicle or brushing against it in the event of a collision	UN regulation No. 26—Rev. 1, No. 61
Protection of pedestrians and other vulnerable road users in collision with vehicles	Global technical regulation No. 9
Installation on vehicles protective devices against underrun of other road users in the event of rear, lateral or frontal collision	UN regulation No. 58—Rev. 3, No. 73—Rev. 1, No. 93

Their content generally corresponds to the regulatory documents discussed above. Properties of anthropomorphic dummies used in crash tests are specified in the next part of CFR (CFR Part 572). The reporting of vehicle manufacturers on the phased implementation of improved safety systems provided for in FMVSS is determined by provisions of the document (CFR Part 585).

Safety of the car occupants in accidents associated with run-off road is provided, along with the properties of the car, by road restraint systems designed to withstand a crash. However, the function of such systems—holding the car on the road—causes additional requirements for minimizing the injuries when hitting a road barrier. These requirements are contained, for example, in the European standards of the EN 1317 series: (EN 1317-1:2010), (EN 1317-2:2010), and in the corresponding to them in the whole, interstate standards used in the Russian Federation: (GOST 33128-2014) and (GOST 33129-2014).

Levels of impact on the car occupants when colliding with a road barrier and consequences of the collision are assessed based on the results of crash tests using the following characteristics (EN 1317-1:2010)[22]:

- Acceleration Severity Index, ASI;
- Theoretical Head Impact Velocity, THIV;
- Post-impact Head Deceleration, PHD;
- Occupant Impact Velocity, OIV;
- Occupant Ridedown Acceleration, ORA;
- Vehicle Cockpit Deformation Index, VCDI (it is modification of the previously proposed Vehicle Interior Deformation Index, VIDI).

ASI is calculated as the maximum of values that characterize acceleration of people sitting near the center of mass of the car at various moments of impact, t_i:

[22] In the Russian Federation, some of them are used according to (GOST 33128-2014).

$$\mathrm{ASI} = \max\left\{\sqrt{\left(\frac{\overline{a}_x}{\hat{a}_x}\right)^2 + \left(\frac{\overline{a}_y}{\hat{a}_y}\right)^2 + \left(\frac{\overline{a}_z}{\hat{a}_z}\right)^2}\,\Big|_{t_i}\right\} \tag{2.3}$$

where $\overline{a}_x, \overline{a}_y, \overline{a}_z$ are components of the acceleration vector averaged over a moving time interval 0.05 s;

$\hat{a}_x$, $\hat{a}_y$, $\hat{a}_z$ are the limit values of accelerations along longitudinal axis of the vehicle, in transverse horizontal direction, and in vertical direction accordingly. For passengers wearing safety belts, these values are: $\hat{a}_x = 12$ g, $\hat{a}_y = 9$ g, $\hat{a}_z = 10$ g.

The greater is ASI, the higher is the risk of injury. In accordance with the Russian (interstate) standard (GOST 33128-2014), the value of this index when hitting a roadside barrier should not exceed:

- 1.0 for passenger cars;
- 1.1 for buses or trucks when using barriers with holding capacity up to and including 450 kJ;
- 1.3 for trucks or articulated heavy goods vehicles when using barriers with holding capacity 500 kJ and more.

VCDI is calculated using the ratios of internal dimensions measured during the tests: after and before impact (EN 1317-1:2010), (EN 1317-2:2010). The Russian equivalent of this index, the coefficient of preservation of internal dimensions, must be at least 0.8–0.9 (GOST 33129-2014).

The tests are carried out on a special test site with recording equipment, using road restraint systems of various types and soils corresponding to the installation site. Speed of the vehicle to be tested depends on its type and mass. For passenger cars, it is in the range from 80 and 100 km/h, for trucks—from 50 to 75 km/h, for buses—from 60 to 72 km/h, for articulated heavy goods vehicles with the mass 35 tons—56 and 60 km/h (GOST 33129-2014). Thus, the conditions of these tests differ from the certification crash tests considered above (UN Vehicle Regulations).

When evaluating the properties of road restraint systems comparatively, or introducing small changes in their design, or considering slight variations of road surface quality or vehicle behavior during an accident, it is allowed to replace tests with computer simulation using software codes that take into account nonlinear structural dynamics (GOST 33128-2014), (GOST 33129-2014).

In addition to mandatory vehicle tests, independent crash tests are also conducted; their results determine ratings of cars in terms of safety. Such independent tests are carried out, for example, by the organization European New Car Assessment Program (Euro NCAP). Its assessments of safety levels are taken into account by both car manufacturers and car consumers.

Compliance with the safety requirements established for the car accidents is provided in the design stage of a vehicle; when designing, the concept of crashworthiness is used, as for aircraft (see Sect. 2.4.2). Implementation of this concept in relation to cars covers giving protective properties to the structure, including those based on irreversible deformations of special energy-absorbing elements, limiting traumatic contacts with equipment and interior of the cabin, using other passive as well as active safety systems, maintaining the possibility of post-crash evacuation of car occupants.

For the design, it is important to determine scenarios of possible accidents. They can be ranked by relative share as follows (Khusainov and Kuzmin 2011):

- frontal collisions including cases of off-center impacts—64%;
- lateral collisions—20%;
- rollovers—10%;
- rear impacts—6%.

Protective properties of the vehicle structure are determined by the strength of main structural elements under impact loads applied during an accident and by characteristics of the zones of programmable irreversible deformations (in which kinetic energy is absorbed). For modern passenger car, the impact loads caused by frontal collision are taken primarily by bumper and the energy-absorbing structure of the front part, including crossbar, crash boxes and spars (see Fig. 2.23).

Crossbar distributes loads on two crash boxes, whose task is to absorb energy in relatively weak impacts. The greatest energy absorption occurs when structural elements lose stability under conditions of plastic flow. To initiate this mechanism of destruction, crash boxes have the corresponding design with instability triggers, for example, initial deviations from the rectilinear form (see Fig. 2.24).

Fig. 2.23 Crash energy-absorbing structure of the front part of a modern passenger car

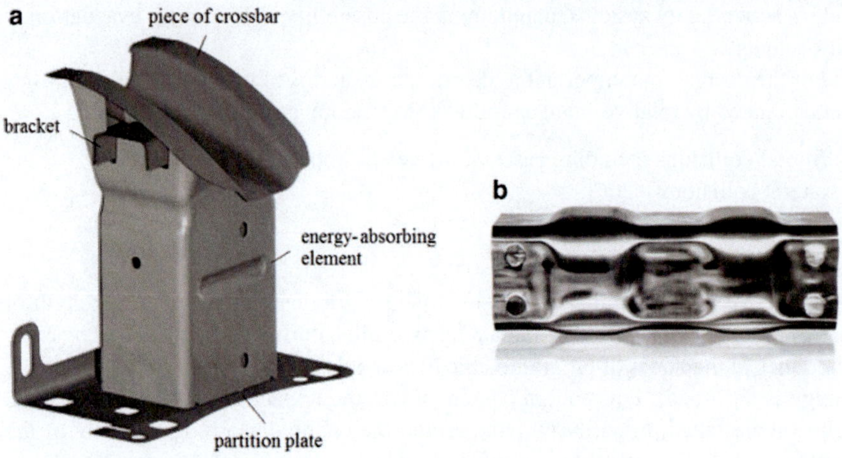

Fig. 2.24 **a** An example of a crash box assembly (Reproduced from Wellkamp and Meywerk (2019). CC BY 4.0), **b** a variant of the energy-absorbing element (Image by PSTproducts GmbH. Reproduced from https://commons.wikimedia.org/w/index.php?curid=39740412. CC BY-SA 3.0)

Severe impacts are extinguished largely by spars, which are designed with folds and stress concentrators to ensure controlled destruction with increased energy absorption. When significant deformations of spars occur, wing reinforcements, also equipped with stress concentrators, come into operation. Part of energy is dissipated by elements of the interior frame. As a result, reduction of impact loads on car occupants is achieved.

To minimize the injuries due to uncontrolled contacts of human body with equipment and interior parts of vehicle (caused by uncontrolled movements during impact), restraint systems: safety belts and inflatable pillows, as well as other passive and active safety devices are used. Here, it should be noted that distinctive features by which safety devices are divided into passive and active ones in automotive industry differ from those that considered in other branches of technology. The author believes that the approach used in nuclear power or in aerospace engineering is generally applicable: the category of passive safety devices includes those that operate on the basis of natural activation, for example, under inertia or gravity forces, without additional energy sources. The active ones act only when they are supplied with energy from any accumulating artificial sources. According to this approach, inertial seat belts (installed on most modern cars) belong to the first category (passive devices), and airbags belong to the second, i.e., to active safety devices.

The inertial belts come into action with a slight delay. Due to this delay, as well as because of gaps and deformations of the belt mechanism elements, movement of a car occupant in the initial phase of impact is possible. Pre-tensioning of the belts minimizes this possibility, but does not completely eliminate it. For additional retention of the driver and passengers from uncontrolled movements, front, side, head-protecting and knee airbags are used (see Fig. 2.25). Each of these airbags is an elastic shell to be filled with air or gas from an inflation device (e.g., pyropatron) triggered by a signal from one or more sensors (usually accelerometers). In addition to the retention function, airbag also redistributes and reduces impact loads acting on human during an accident.

There are seat belt systems containing an inflatable element built into the chest belt to reduce possibility of chest injuries, as well as airbags for motorcyclists.

However, airbags are ambiguous safety ensuring systems: sometimes, for example, without seat belts, they can cause harm to occupant's health. The fact is that the elastic shell (the volume of which for the front passenger airbag is about 130 liters) is inflated and unfolded with speed up to 300 km/h. To avoid injury in such a rapid process, the distance between an occupant and airbag at the triggering time moment should be at least 250 mm.

The driver's front airbag has usually a smaller volume—50 to 80 liters, and the impact energy is absorbed partly by the steering wheel. The injury safety of the steering wheel is ensured by a slotted connection of steering elements (with slippage), a deformable shaft in the form of a bellows, a perforated sleeve, etc.

To dangerous consequences of a car accident relate

- fire due to a short circuit in the electrical wiring, a fuel spill, or the ingress of combustible materials into the exhaust pipe;
- filling the car that has fallen into a pond with water.

Evacuation of victims of accident can be complicated by overturning the car, jamming the doors, the need to release people from their seat belts, and other similar circumstances. The possibility of timely leaving the car in all threatening conditions should be supported by appropriate constructive measures. In order to save the survivors, the car equipment may include a system for notifying emergency services about the location and nature of accident. Besides, event data recorders (EDRS) are

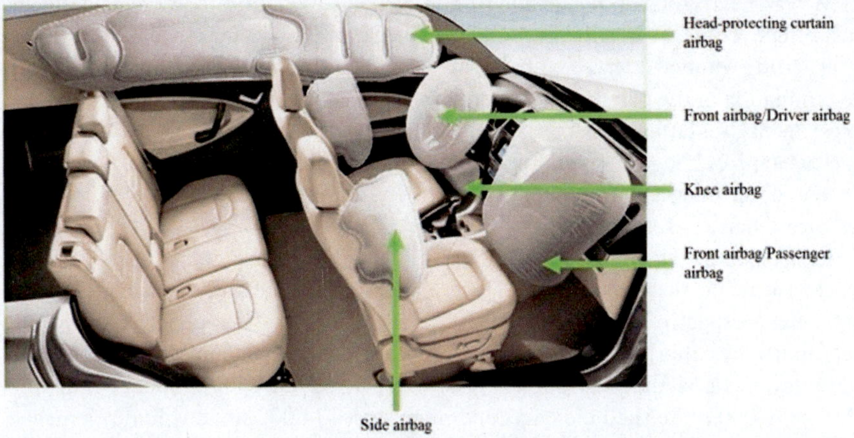

Fig. 2.25 Airbags. Reproduced from https://techautoport.ru/wp-content/uploads/2019/08/2019-08-06_15-52-55.png

installed in modern cars, which allow analyzing the behavior of their systems at all stages of an accident.

The design of vehicle in accordance with the crashworthiness concept is based on analytical simulation of crash impacts, which is carried out along with crash tests, and sometimes as an alternative to them. In many cases, simulation of the vehicle structural dynamics and simulation of the behavior of its occupants during impact are performed separately. As result of the first simulation carried out with an analytical model of the car structure only, acceleration pulses are determined, and then they are used as input data for analyzing dynamic reaction of the human body modeled independently (Du Bois et al. 2000; Ambrosio and Silva 2001). However, for some crash situations including, for example, car rollover, separate modeling and simulation of behavior of these mechanical objects (vehicle structure and occupants) are not applicable. Significant inaccuracies also arise when studying crash impact of a small car, the mass of which is comparable to the mass of an occupant. In such cases, analysts have to use combined models that take into account interaction of passengers with the vehicle structure, influence of the suspension of the wheels and characteristics of the ground surface.

For analytical modeling, the following methods are applied: the lumped parameters method (LPM), the macro element method (MEM) and the already referred FEM. Models with lumped (concentrated) inertia and stiffness parameters—masses and nonlinear springs—have been repeatedly mentioned above (see Sects. 2.4.4 and 2.4.5). LPM was probably the first method used to analyze dynamic behavior in crash impact of both car structure (Kamal 1970; Kamal and Lin 1982; Ambrósio and Pereira 1997), and human body (Laananen 1974; Huston et al. 1976; Lyakhovenko and Merkur'ev 1995). This method is based on the assumption that a deformable object can be represented by a number of rigid segments each characterized by mass and moments of inertia, and bonds with elastic, plastic and viscous properties. Special nonlinear connections can also describe variable contacts. Subsequently, LPM was nearly ousted by detailed finite element modeling; however, due to its cost-effectiveness, it is still used today (Kostek and Aleksandrowicz 2017; Deac et al. 2017; Munyazikwiye et al. 2018). Its limitation is the need for preliminary determination of the parameters of nonlinear springs. They are determined either by experiments, including crash tests, or by calculations based on FEM.

This approach is presented, for example, in the publication Munyazikwiye et al. (2018), describing a LPM simulation of a Ford *Taurus* crash test: frontal collision with rigid barrier (see Fig. 2.26).

Resistance of the front part of the car under impact loads was simulated using piecewise linear dependences of elastic and viscous forces on displacements and velocities, respectively (a Kelvin model). Parameters of these dependencies were determined by calibration based on the results of a crash test at impact speed 56 km/h and relevant FEM analysis. The analysis was carried out with a model containing about 839,000 finite shell elements and more than 134,000 finite volume elements, using the LS-DYNA program.

The calibrated LPM model was used then in simulations of collisions at other impact speed values (40, 48, 64 and 72 km/h) to calculate accelerations, ASI and

Fig. 2.26 Crash test of a Ford *Taurus* (2004 model year): frontal collision with rigid barrier at 56 km/h. Reproduced from Munyazikwiye et al. (2018)

the maximum deformation of the front part of the car relative to the center of mass position, Δ_{max}. These characteristics were compared with experimental data and the results of FEM analyses. The time histories of kinematic parameters, obtained by simulations, corresponded on their trends to those registered in experiments, but still differed from them (when calculating both by LPM and by FEM). Duration of the pulse of acceleration was practically independent of the impact speed and was 0.12–0.13 s. According to the calculation using LPM, in the case of calibration by crash test data, ASI was equal to 2.6, and the value of Δ_{max} was approximately 0.751 m (at 0.0738 s); in the experiment: ASI = 2.5; Δ_{max} = 0.755 m (at 0.0723 s). After calibration of the LPM model by the results of FEM analysis, the calculation output gave ASI = 2 (similar to the finite element result); the difference between the values of Δ_{max} obtained using the LPM and FEM models did not exceed 3 cm for all considered values of the impact speed.

The calibration time was 15–30 min; the calculation took no more than 20 s. Analysis using finite element model lasted up to two days, in addition, an indefinite time was spent on building the model itself. During FEM analyses, it was necessary to control and limit the emerging non-physical effects in the form of so-called "hourglass modes" caused by significant changes in geometry of elements due to large displacements.

In MEM, relative displacements of parts of the structure that are being destroyed during crash impact (crash boxes, spars, crushable steering elements, etc.) are approximated by non-analytical functions obtained on the basis of relevant experiments (Du Bois et al. 2000; Abramowicz 2001). This approach does not have the universality of the finite element modeling based on polynomial shape functions, but, like LPM, it is very efficient in terms of calculation time.

The so named *meshfree methods* that appeared later can be regarded as analogous to MEM; they make it possible to avoid computational problems caused by significant distortions of finite element mesh at large displacements (the main of these problems are "hourglass modes" and instability). Meshfree methods are based on construction of kernel functions for bounded spatial regions by splines approximating displacements of particles within them, and are kindred in this respect to the boundary element method. These methods (implemented, among other numerical tools, in LS-DYNA) are more expensive in terms of computation time than FEM. They are used to model certain parts of the vehicle structure that are subject to the most severe deformations during crash impact, in combination with the discretization of the remaining parts by finite elements (Wang et al. 2006).

Combined modeling by finite elements and macroelements with non-traditional approximations based on empirical and analytical dependencies, as well as reduction algorithms, was proposed and discussed in the author's works (Spirochkin 1993; Spirochkin 1994; Spirochkin 1998). The preference of special shape functions other than Hermite polynomials for structural dynamics problems was substantiated. However, such research and development did not go further largely due to the promotion of foreign software codes in Russia, such as LS-DYNA, which are mentioned even in the interstate standards (GOST 33128-2014), (GOST 33129-2014).

The use of FEM as a convenient and reliable method for modeling the structure and equipment of a car, as well as its occupants over the last decades was being accompanied by a continuous complication of the built models. While in 1985, a VW Polo model has been reported to contain 2272 shell finite elements and 106 beam finite elements (Du Bois et al. 2000), three decades later, models with almost 7 million elements were being used in "virtual" crash tests of General Motors vehicles, and this number more than tripled over the previous 5 years (Golson 2015). Each of these virtual tests was being performed on mainframes of the company's data center in Warren, Michigan, during the night.

The current state of FEM modeling can be illustrated, for example, by the report Singh et al. (2018), in which computer analyses of crash tests of a Honda *Accord* car using LS-DYNA are compared with registered data of the tests themselves. Figures 2.27, 2.28 and 2.29, borrowed from this report, show:

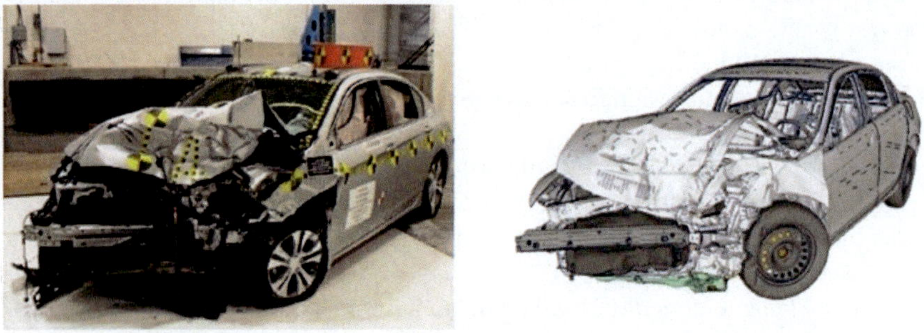

a b

Fig. 2.27 **a** General view of a 2014 model year Honda *Accord* and **b** its analytical (FEM) model after oblique frontal impact into a rigid barrier at 90 km/h. Reproduced from Singh et al. (2018)

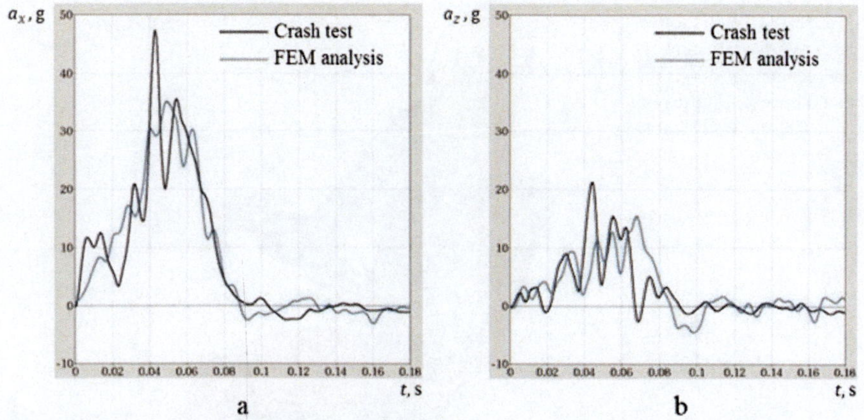

Fig. 2.28 **a** Longitudinal and **b** transversal accelerations versus time near the center of mass of a Honda *Accord* during oblique frontal impact into a rigid barrier at 90 km/h. Reproduced from Singh et al. (2018)

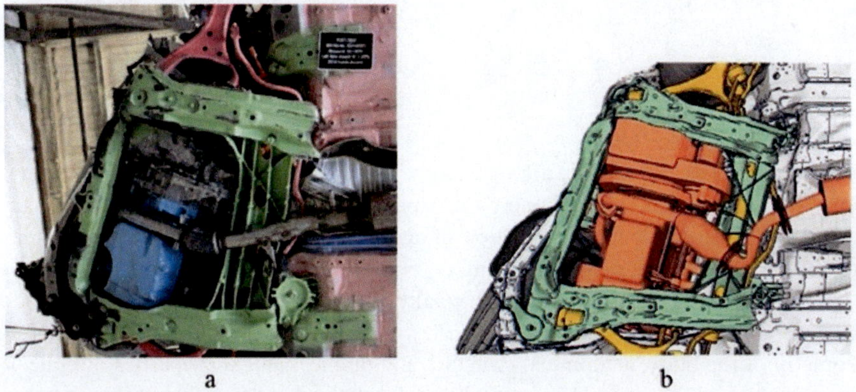

Fig. 2.29 Residual deformations of a Honda *Accord* front part after oblique frontal impact into a rigid barrier at 90 km/h: **a** post-crash photo, **b** FEM analysis result. Reproduced from Singh et al. (2018)

- general view of the car and its analytical (FEM) model after an oblique frontal impact into a rigid barrier at a speed of 90 km/h;
- longitudinal and transversal accelerations during impact;
- deformed front part of the car (bottom view).

The presented results of the analysis correspond, as it can be seen, in general to the data obtained during the crash test, at the same time, there are differences. The last ones manifest themselves both in accelerations (see Fig. 2.28) and in residual deformations—for example, in the greater skew-symmetry of the deformed front part of the car and in the shape of plastically deformed exhaust pipe (Fig. 2.29).

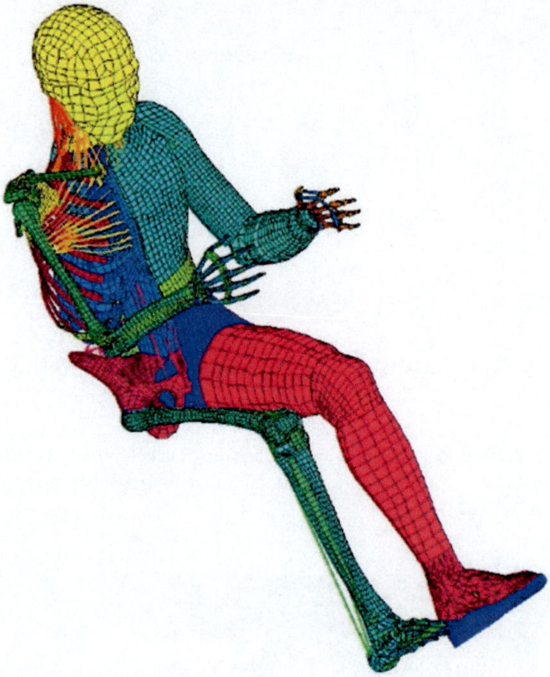

Fig. 2.30 THUMS (some tissues removed to show finite element mesh on skeletal bones). Courtesy of Toyota Central R&D Labs, Inc.

Most likely, these differences are caused by inaccuracies in modeling the mechanisms of structural failure and energy dissipation, as well as insufficiently precise consideration of the contact interaction of aggregates and equipment in the engine compartment of the car when analyzing deformation processes.

Modeling of the human body by LPM, already mentioned, has been widely used until the beginning of the XXI century. FEM models came later and approximated the geometry of the body, anatomical structure and biomechanical properties of bones, tissues and internal organs (Haug et al. 2004). From these models, the most used until now is Total Human Model for Safety (THUMS), designed for LS-DYNA, which represents a 50-percentile[23] male dummy. The initial version of THUMS by Toyota Motor Corporation and its research center Toyota Central R&D Labs has appeared around 2000. The article Oshita et al. (2002) describes one of the first examples of application of THUMS (see Fig. 2.30) for the analysis of safety in a car accident.

Then, this model was being updated with a certain regularity depending on the refinement of biomechanical data. In Version 3, approximately 150,000 finite elements have been used (Vázquez 2014). In Version 4, the modeling of the chest organs has become more detailed and scaling has been implemented, corresponding to the 5-percentile female, 50- and 95-percentile male dummies, as well as children 3, 6 and 10 years old. In Version 5.0, issued in 2015, muscle modeling has been added. This addition allowed analysts to take into account the posture of a driver or

[23] Corresponding to 50% probability of not exceeding biomechanical parameters.

passenger when he or she realizes that a collision will occur and instinctively prepares for it. The latest development is Version 6.0, introduced in 2019. It combines in one software package tools for modeling both pre-crash changes in occupant posture and injuries during collision (Clifford 2020).

In Europe, in 1999–2001, as a result of cooperation of six automotive companies, a number of software manufacturers, research organizations and universities, a similar model HUMOS (Human Model for Safety) has been developed. Its HUMOS2 modification provides refined modeling of joints, ligaments and internal organs, as well as scaling that corresponds to the above percentile variants. This model is applicable when analyzing dynamic reaction of the human body during crash impact by the program codes MADYMO, RADIOSS or PAM-CRASH (Vezin and Verriest 2005).

From recent achievements, R&D of the Global Human Body Models Consortium (GHBMC) shall be mentioned: a very detailed 50-percentile male model M50-O, containing 2.2 million finite elements, and its simplified version M50-OS, in which the number of elements is reduced to 354,000 (Schwartz et al. 2015). These models are intended for use with the LS-DYNA program code.

Biomechanical properties of bones, organs and soft tissues to be assigned for the models, discussed above, are determined experimentally; methods for determination and the data themselves are presented, for example, in publications King (2000, 2001), Haug et al. (2004) and Vezin and Verriest (2005).

Detailed finite element models allow analysts to take into account the properties of certain parts of the human body, determine deformations, stresses and accelerations in them, as well as to understand damage mechanisms, provided the description of these properties and mechanisms is correct. The price for accuracy is significant costs including the time spent for numerical modeling and simulation. In a number of cases, the LPM modeling remains preferable, especially when a quick assessment of kinematics and general dynamic reaction of the human body is required.

When designing a wheeled vehicle according to the crashworthiness concept, data on behavior of its structure and systems, as well as on injuries of occupants in real road accidents, are of considerable value. These data differ depending on the types of vehicles. Thus, statistics on bus accidents can be found, for example, in the publication Matolcsy (2007) and the report Olivares (2012), on heavy trucks—in the reports NHTSA (2015) and Woodrooffe and Blower (2015), and so on. The existing databases on accidents and crash tests, as well as methodological aspects related to extracting information from them for correctly establishing a connection between conditions of crash impacts and injuries, are discussed in the article Vangi et al. (2018).

2.4.7 Rail Vehicle Crashes

It is assumed that the accident rate for railway transport (railway trains, metro trains, urban and suburban trams) is lower compared to aviation and road transport (Pereira 2006). However, accidents cannot be completely excluded, and over the past decades, the topic of passive safety has become increasingly important in the railway industry. In order to reduce injuries to passengers and crews, to ensure their survival in possible crashes, research and development on crashworthiness of railway vehicles is carried out over the world. As in the case of aircraft or car, safely crushable structure of rail vehicle represents the last level of occupant protection when all possibilities for preventing an accident have been exhausted.

Analysis of railway crashes shows that most fatalities and serious injuries occur in end-on collisions accompanied by overturning of coaches (Seitzberger et al. 2009). Accordingly, to increase chances of survival and reduce injuries, the main attention in the design of railway vehicles is paid to the ends of their elements (locomotives and coaches) where it is possible to ensure absorption of impact energy and prevent overriding over each other, leading to overturning.

In the EU countries, the regulatory document (EN 15227:2008) is in effect; its requirements are mandatory for all new railway vehicles since 2012. The provisions of this document relate to:

- categorization of vehicle structures by crashworthiness;
- design collision scenarios;
- constructive measures of passive safety taking into account necessary survival space, tolerability of acceleration pulses, evacuation opportunities and so on;
- validation of crashworthiness properties.

For railway vehicles, the following design scenarios are established: train-to-train front-end collision or train collision with a standing 80-ton freight wagon at 36 km/h, train collision with a car on road-railway crossing at 110 km/h, etc. Due to high cost, full-size certification crash tests of coaches and locomotives are not performed—instead, computer simulation is carried out, including that based on FEM models, and only certain vehicle parts are tested. The results of relevant computer simulation and test should not differ by more than 10%.

Requirements for crashworthiness and crash energy management have influenced significantly on the design of the head-parts of high-speed trains. These parts began to be designed as separate modules in the form of cabins equipped with crushable (energy-absorbing) structural elements and anti-climbers (Fig. 2.31).

In the USA, comparable requirements are established by the Passenger Equipment Safety Standards (CFR Part 238), which are administered by the Federal Railroad Administration (FRA). In development of these standards, as well as relevant rules and recommendations, a number of organizations are involved including the American Public Transportation Association (APTA). The APTA's approaches to ensuring crashworthiness of passenger coach structures, which are included in standardization documents of this organization, supplement the FRA requirements and are recognized by this regulator, are given in the article Tyrell (2002).

In Russia, when designing, manufacturing and certifying railway rolling stock for 1520 mm track gauge, it is necessary to comply with the requirements of the following mandatory documents:

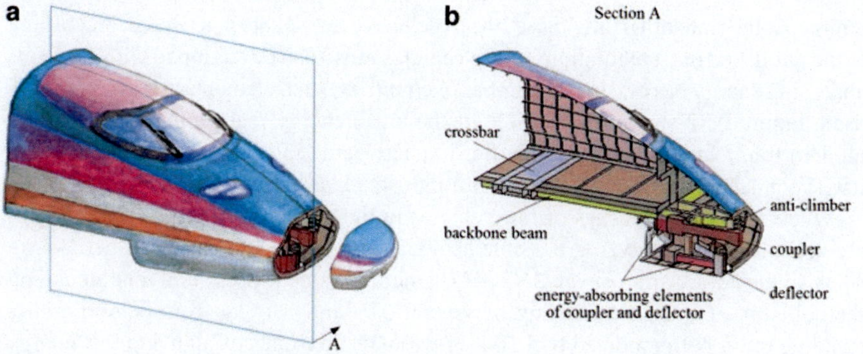

Fig. 2.31 **a** General view and **b** load bearing structure of a high-speed train cabin

- Technical Regulations (TR TS 001/2011) applying to rail vehicles with a speed of no more than 200 km/h and
- Technical Regulations (TR TS 002/2011) related to high-speed rail transport (more than 200 km/h).

According to these documents, designer must provide for "emergency crash systems" to protect passengers and personnel in the event of collision and (or) derailment of the train, as well as ensure exclusion of injuries due to the destruction of glazing under deformations during impact. In order to comply on a voluntary basis with the requirements of the second mentioned technical regulations, the interstate standard (GOST 32410-2013) is used. It complies with the European regulatory document (EN 15227:2008) in principles of ensuring passive safety by energy-absorbing elements that limit the levels of accelerations and forces during crash impact.

In collisions with obstacles on scenarios similar to the listed above, energy absorption by these elements should provide an average value of longitudinal acceleration no more than 5 g in the middle of a coach body at floor level. Residual deformations of the body in the longitudinal direction and changes in the dimensions along diagonals of the door and window openings should not exceed 1%. Elements (devices) absorbing crash energy should be placed in front and end parts of locomotives, in cantilever parts of passenger coaches, and in motor-car rolling stock—in front part of the head unit and, if necessary, between units.

Verification of characteristics of such devices (energy absorption capacity, deformation diagrams, etc.) is based on full-size tests. According to test data, mathematical models of their nonlinear deformation should be developed which in the case of acceptability can be used in the calculations of rolling stock collision processes. In order to ensure the correct simulation of force transmission and propagation of deformations, the standard regulates the three-dimensional modeling of cabin modules and the main load bearing structural elements of locomotives and coaches equipped with energy-absorbing devices (similar to that shown in Fig. 2.31).

Safety of metro trains and light rail vehicles, including trams, should be governed in the Russian Federation by relevant technical regulations of the Eurasian Economic

Union. At the time of writing these lines, they have not yet been accepted. According to the published draft regulations, these vehicles must also be equipped with elements absorbing energy in a crash. In a number of countries, such elements are already being used. Figure 2.32 shows an example of the crash energy management system of a modern tram, which consist of a buffer impact beam (an energy-absorbing element itself) connected to the vehicle body through shock absorbers.

When simulating crashes of rail vehicles in the design stage, the same methods are used, as when analyzing possible accident of a developed car (see Sect. 2.4.6). Thus, in the book Ambrosio et al. (2001), a methodology is presented for calculating the collision of a train consisting of several cars and coming from behind with a standing train, both modeled by LPM. Previously, the zones of high and low energy absorption were identified: the first ones located in the head part of the coming locomotive and the tail part of the last standing coach, the second ones—in areas of each coach couplers. Each coach was modeled by 8 rigid bodies with lumped masses (representing body parts, couplers, wheels); coach suspension was approximated by springs and dampers; interaction of wheels with rails was described using nonlinear elastic-viscous elements with friction. In order to simulate the behavior of deformable parts corresponding to the above-mentioned energy absorption zones, elastic–plastic

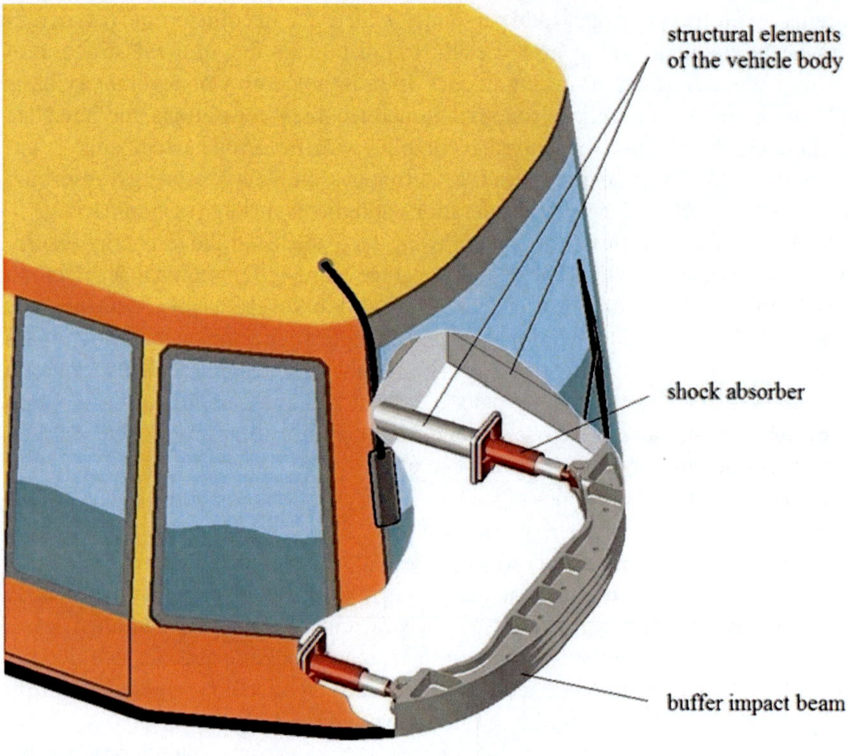

Fig. 2.32 Crash energy management system of a modern tram

diagrams were put into the calculation. Functioning of anti-climbers (positioned in the head of the coming locomotive and couplers of coaches) was imitated as contact with stiction.

The use of FEM allows analysts to take into account, when calculating the collisions of rail vehicles, the following factors (Kirkpatrick et al. 2001):

- nonlinear behavior of materials (what is more fundamental and precise than the simulation of behavior of deformable parts);
- large displacements and rotations of structural elements;
- contact interactions between them;
- welds and fasteners as well as effects of cracks and damages.

The set of crash analysis tools, which comprised at the end of the XX century the above-mentioned codes PAM-CRASH, RADIOSS, MSC.Dytran and LS-DYNA, was supplemented later by the FEM programs ANSYS and ABAQUS. A typical numerical model of a railway vehicle at the beginning of the 2000s included 50,000 to 250,000 finite elements.

As a result of studies conducted in Europe for regional trains,[24] required characteristics of energy absorption zones (corresponding to average longitudinal acceleration no more than 5 g) were established: energy absorption capacity, E, and maximum deformation, Δ_{max}. Their values for high-energy absorption zones are: $E = 4.6$ MJ; $\Delta_{max} = 1.8$ m, and for low energy absorption zones: $E = 0.7$ MJ; $\Delta_{max} = 0.7$ m (Pereira 2006). Based on these data, end parts of coaches of a regional train were designed. Crash tests were simulated using finite element software tools ANSYS, RADIOSS and PAM-CRASH; calculation data were compared with the test results. Similar work was carried out for trams. The required crash energy absorption for them estimated as 135 kJ (Pereira 2006).

Currently, there are no agreed requirements for ensuring safety of passengers, train or tram drivers, taking into account their movements inside the vehicle during crash and contacts with interior elements, except for the obvious provision to exclude sharp edges of elements. Studies in this direction were being performed (Kirkpatrick et al. 2001; Pereira 2006) and are in progress now; in them, mathematical models of vehicle occupants, similar to those described in Sect. 2.4.6, are used.

The features of the processes of destruction in rail vehicles in collisions, determined by weld failures or other initial technological imperfections (cracks, damages of geometry), began to manifest themselves significantly in monocoque aluminum structures of high-speed trains and metro trains. For example, a phenomenon of catastrophic "weld unzipping" was discovered. It was occurring due to dynamic plastic rupture of the metal in the weld area or in the heat-affected zone (Kotsikos 2010). These features led to the need for:

- fracture mechanics tests and impact tests to clarify mechanical properties and to trace propagation of impact loads in structural elements fabricated by extrusion and welding;

---

24 Their speed does not exceed 200 km/h.

- extensions of the set of strength criteria;
- develop the finite element modeling of weld failures and other imperfections;
- reinforcement of the structure in the weld vicinity.

Procedures for assessing crashworthiness of railway passenger coaches in the case of overturning are not under regulation now. The regulatory documents discussed above imply indirectly that energy-absorbing elements and anti-climbers minimize the risk of overturning. However, this risk remains, and considering it, provisions of the UN Regulation No. 66, which apply to large-sized wheeled vehicles (see Table 2.4 in Sect. 2.4.6), are sometimes used. Such approach is described, for example, in the article Baykasoğlu et al. (2013). Computer simulation of the behavior of a coach when overturning on hard ground was performed using the finite element code ABAQUS. The developed FEM model was verified by the data of static and dynamic modal tests. The simulation allowed the researchers to find that the space inside the coach meets requirements of the mentioned regulation during and after overturning.

Recently, in computer simulations of the railway crashes accompanied by large relative displacements of structural components, as in relevant studies of car accidents, coupled models (which include FEM discretization and meshfree approximation) are being applied. Thus, using the LS-DYNA version that supports this combined methodology, a collision of the leading coach of a high-speed train with a rigid obstacle was analyzed (Tang et al. 2016). The meshfree method was used to simulate energy-absorbing devices, backbone beam and a number of other structural elements subject to severe deformation; the remaining structural components were modeled by FEM, the total number of finite elements was about 254,000. Simulation of the collision process (at a speed of 80 km/h) during a real time interval 0.3 s took more than 30 h. Duration of the impact pulse (determined by the corresponding changes in forces and energy absorption in structural components) was approximately 0.25 s. The modeling inaccuracy in terms of the amount of absorbed energy was estimated as 0.72%, but for internal forces it reached 20%.

2.5 People Under Conditions of Malicious Actions

2.5.1 Protection Against Malicious Actions

Let us move on to the last of the identified categories of roles that characterize human interaction with a technical system. It consists of two roles, or, more precisely, sub-roles: *attacker* and *potential victim* (see Sect. 2.1).

Attackers, including terrorists, can cause harm through their deliberate actions aimed at destroying a technical system saturated with energy and (or) releasing dangerous technological substances contained in a system into the environment. Risk of malicious actions is taken into account in nuclear industry. The main approach to minimizing this risk is the physical protection of nuclear facilities. National regulators in different countries have developed rules to create and support the physical protection systems for nuclear materials, nuclear power plants and other nuclear facilities in which these materials are used.

In Russian nuclear power industry, some of these rules are collected in the regulatory document (NP-034-15). It refers to radioactive substances, radiation sources and storage facilities, with the exception of certain facilities and activities, for example, transportation of radioactive substances and waste. The purpose of the physical protection system is to counteract intruders defined by the "violators' model". Such the model is "a set of data about the number, equipment, readiness, awareness and tactics of actions of violators, their motivation and the goals they pursue". The physical protection system should cover organizational measures, technical means and personnel. Its tasks include:

- prevention of deliberate actions that can lead to an emergency situation and create a radiation threat (theft of radioactive substances, unauthorized access to the protected area of nuclear facility, failure of protection equipment, etc.);
- timely detection of unauthorized actions;
- hindering the progress of violators;
- responding to unauthorized actions and neutralizing violators.

Deliberate unauthorized threatening actions of intruders are characterized as "sabotage". The term "protected area" means an area where access is restricted and controlled; it is surrounded by physical barriers that are constantly guarded and monitored. Depending on the degree of danger attributed to the consequences of sabotage, different levels of the physical protection are established. In practice, differentiation is based on the recommendations given in the safety guideline (RB-130-17).

The rules of physical protection during the transportation of radioactive substances and radiation sources by all modes of transport on the territory of the Russian Federation are established in the regulatory document (NP-073-11). The purpose of this kind of protection, as in the previous case, is to prevent the theft of these substances and sources, as well as sabotage in their relation, which can lead to unacceptable radiation exposure on people and the environment. To prevent unauthorized actions, access to vehicles, cargo transported in them and information about physical protection is restricted. In addition, protection equipment is used corresponding to the established "violators' model", the population is informed about cargo protection and responsibility for unauthorized actions, and "other measures are accomplished that contribute to reducing the confidence of violators in the successful implementation of threats".

In the nuclear power industry of the USA, the physical protection of nuclear facilities is regulated by the rules established by the Nuclear Regulatory Commission (NRC)—see (CFR Part 73). The applicant for the license to operate a nuclear facility (the licensee) must provide for protection measures against acts of radiological sabotage and prevent the theft or diversion of special nuclear material. Such threats should form the design basis[25] for the physical protection system to be used.

By "radiological sabotage", NRC understands, first of all,—see (CFR Part 73, § 73.1)—"a determined violent external assault, attack by stealth, or deceptive actions, including diversionary actions, by an adversary force"—one or more individuals, groups or their combination. The adversary force may consists of "well-trained (including military training and skills) and dedicated individuals, willing to kill or be killed, with sufficient knowledge to identify specific equipment or locations necessary for a successful attack". They can have (the attributes of the aforementioned "violators' model" are actually listed below):

- "active (e.g., facilitate entrance and exit, disable alarms and communications, participate in violent attack) or passive (e.g., provide information)" equipment, or both, as well as "knowledgeable inside assistance";
- "suitable weapons, including hand-held automatic weapons, equipped with silencers and having effective long range accuracy";
- "hand-carried equipment, including incapacitating agents and explosives for use as tools of entry or for otherwise destroying reactor, facility, transporter, or container integrity or features of the safeguards system";
- "land and water vehicles, which could be used for transporting personnel and their hand-carried equipment to the proximity of vital areas".

Radiological sabotage also covers (CFR Part 73, § 73.1):

- an "internal threat";
- a "bomb assault" using a land or water vehicle, "which may be coordinated with an external assault";
- a "cyber-attack".

The theft or diversion of special nuclear material in an amount sufficient for the production of nuclear fuel is characterized similarly to radiological sabotage.

According to the requirements of NRC, a fixed site physical protection system must (CFR Part 73, § 73.45):

(1) "prevent unauthorized access of persons, vehicles and materials into material access areas and vital areas[26]";
(2) permit within these areas "only authorized activities";
(3) permit within material access areas "only authorized placement and movement of strategic special nuclear material[27]";

[25] The term "design basis" means information that defines functions of a designed system as well as the specific values or range of values that the project should meet. When establishing these functions and values, all possible impacts of natural and anthropogenic origin should be taken into account.

[26] Material access area is a location which contains special nuclear material and is equipped with physical barriers. Vital area means any area which contains vital equipment.

[27] Strategic special nuclear material means uranium-235, uranium-233, or plutonium.

(4) "permit removal of only authorized and confirmed forms and amounts of strategic special nuclear material from material access areas";
(5) "provide for authorized access and assure detection of and response to unauthorized penetrations of the protected area";
(6) "assure that the five capabilities described" above "are achieved and that adversary forces will be engaged and impeded until offsite assistance forces arrive".

The components of the physical protection system that implement these functions are (CFR Part 73, § 73.46):

- an established by the licensee security organization, including guards;
- physical barriers that separate protected areas from the surrounding space (fences, walls, ceilings and floors of buildings, etc.);
- access control, detection, surveillance and alarm subsystems and procedures;
- communication devices for guards or armed response personal;
- test and maintenance programs for intrusion alarms, emergency exit alarms, communications devices, physical barriers, and other physical protection related equipment;
- contingency and response plans and procedures;
- schedule for revisions to physical protection plans.

The functioning of the physical protection system shall provide for restrictions on public information about the protection measures taken. The restrictions relate to (CFR Part 73, § 73.22):

- "the composite physical security plan for the facility or site";
- "site-specific drawings, diagrams, sketches, or maps that substantially represent the final design features of the physical security system not easily discernible by members of the public";
- "alarm system layouts showing the location of intrusion detection devices, alarm assessment equipment, alarm system wiring, emergency power sources for security equipment, and duress alarms not easily discernible by members of the public";
- "physical security orders and procedures issued by the licensee for members of the security organization detailing duress codes, patrol routes and schedules, or responses to security contingency events", etc.

Physical protection during transportation of nuclear materials is also characterized by restrictions on public information. They concern:

- "the composite physical security plan for transportation";
- "schedules and itineraries for specific shipments of source material, byproduct material, high-level nuclear waste, or irradiated reactor fuel";
- "vehicle immobilization features, intrusion alarm devices, and communications systems";
- "arrangements with and capabilities of local police response forces, and locations of safe havens identified along the transportation route", and so on.

Relatively new are the requirements of NRC for the licensee to protect a nuclear power plant from cyber-attacks. In accordance with these requirements (CFR Part 73, § 73.54), protective measures in relation to digital computer and communication systems and networks shall be provided.

As part of the physical protection system, the licensee is also required to develop and implement a program to mitigate the consequences of insider activity. This program (CFR Part 73, § 73.55) "must monitor the initial and continuing trustworthiness and reliability of individuals granted or retaining unescorted access authorization to a protected or vital area, and implement defense-in-depth methodologies to minimize the potential for an insider to adversely affect, either directly or indirectly, the licensee's capability to prevent significant core damage and spent fuel sabotage".

* * *

The organization of protection against malicious actions is also typical for other industries where potentially dangerous technical systems are used. At the same time, of course, there are differences, including in terminology. For example, the term "physical protection" is not used in aviation, but within the framework of aviation safety, measures are implemented in many ways similar to the physical protection of nuclear power plants.

According to the Russian *Air Code* (see Chapter XII), aviation safety is "the state of protection of aviation from illegal interference" in its activities. Illegal interference means, among others, unlawful actions that threaten safety, "resulting in accidents with people, material damage, seizure or hijacking of an airplane". The functions on ensuring aviation safety include:

(1) prevention of extraneous individuals' access to the controlled areas of airports (airfields);
(2) protection and safeguard of airplanes in the parking lots from extraneous individuals entering them;
(3) exclusion of the possibility for unauthorized transportation of weapons, ammunition, explosives, radioactive, toxic, flammable substances and other similar sources of danger in an aircraft;
(4) special precautions when allowing their transportation;
(5) pre-flight, and if required, post-flight inspection of passengers;
(6) implementation of measures to counteract acts of illegal interference in aviation activities, including with the involvement of law enforcement agencies;
(7) exclusion of the possibility of unauthorized access to unmanned aerial vehicles (UAVs).

These functions are carried out by airport (airfield) aviation security services, departmental security units (subordinate to the federal executive authority in the field of transport), aviation security services of organizations operating airplanes, as well as authorized bodies (endowed with this right by federal laws).

Regarding technical means of inspection intended to detect weapons, explosives and other substances and items prohibited for transportation by airplanes in the Russian Federation, the national standards (GOST R 55249-2012) and (GOST R 58777-2019) are applied.

The term "design basis" (typical for nuclear power industry) is not used in aviation, but the airworthiness standards for transport category airplanes, e.g., (CFR Part 25) or (AP-25 2009), provide for ensuring the safety of the flight crew in the aircraft design stage, based on specified parameters of anthropogenic malicious effects. Thus, the flight crew compartment must be protected from violent intrusion of extraneous individuals who are not included in the crew. According to relevant provisions of the above standards, the bulkhead, door and its installation elements must:

- withstand impacts "in critical places" (AP-25 2009, § 25.795) with energy 300 J or static tensile load 1113 N "on accessible handholds, including the doorknob or handle" (CFR Part 25, § 25.795);
- resist penetration of small arms bullets and fragments of explosive devices with specified parameters.

2.5.2 Early Detection of Violators

Detection of violators before they commit criminal actions (terroristic act, sabotage, etc.) is another opportunity to ensure safety, taking into account the malicious manifestations of human factors. It can be used in combination with a physical protection system, to identify and neutralize relevant threat in an early stage, when there is no contact of attackers with this system yet, or independently.

There are two main approaches to early detection. The first one, called profiling, consists in performing the detection function by specially trained personnel. Within this approach, psychological methods are applied to assess the personality and intentions of an individual based of particular signs, characteristics of appearance, non-verbal and verbal behavior. Profiling in the form of observation and interviews has been first introduced by the Israeli airline *El Al* to identify potentially dangerous persons among air passengers. As a result, this company acquired the status of one of the safest in the world. According to the latest edition of the *Rules for conducting pre-flight and post-flight inspections*[28] and the *Rules for conducting inspection, additional inspection, re-inspection in order to ensure transport safety*,[29] profiling has now become a practice in the Russian transport sector.

The second approach is instrumental: it is based on the use of technical devices for recognizing individuals, monitoring their behavior and predicting intentions. The most common technology is facial recognition by identifying digital images obtained

[28] Approved by the Directive of the Ministry of Transport of the Russian Federation No. 104 of 25.07.2007 (Ed. of 19.08.2019).

[29] Approved by the Directive of the Ministry of Transport of the Russian Federation No. 227 of 23.07.2015.

from surveillance cameras installed in places of mass visits of people, including transport infrastructure facilities. To identify violators, a comparison of characteristic facial features with biometric indicators stored in databases is used. To eliminate the influence of glasses, hats and make-up, faces are photographed in infrared rays.

Some information about facial recognition and the place of this technique among the methods of biometric identification of people with a criminal past is presented on the website of the U.S. Federal Bureau of Investigation (FBI 2020). The state of R&D in this area performed in the world until the end of the first decade of the XXI century is discussed in the publication Crawford (2011).

The relevant work that is being carried out in Russia can be judged, for example, by the dissertations (Timoshenko 2014; Sorokin 2017). In the first of them, the existing methods of facial recognition on photographic images are considered and a new method is proposed that is applicable under uncontrolled shooting conditions (poor lighting, unfavorable position of the head in relation to the camera, aging, facial distortions, etc.). Potential areas of its application are surveillance systems in public places: in the subway, at train stations, at airports. The recognition accuracy for the proposed method is at the level of 98%.

The second thesis is devoted to information and analytical support of security management in places of mass stay of people based on the identification of violators. The movements of each violator in the field covered by the surveillance cameras are described by finite homogeneous Markov chain; i.e., it is a sequence of random events, properties of which include a finite or countable infinite number of outcomes and independence of the future from the past. Based on the estimated probability of detecting a violator in a certain place, a probabilistic graph of his movement routes is constructed and, taking into account the location of security service officers, coordination of forces and means to counter malicious actions is ensured. An improved hybrid identification algorithm based on facial biometrics is proposed, which differs from prototypes in features of data registration and processing. The results of its application demonstrated an average accuracy about 98.2%; the identification time was 0.377 s.

In the Russian Federation, the national standard (GOST R 58776-2019) for monitoring human behavior and predicting intentions of people has recently been put into effect. The application area of this standard is vaguely defined, but it can be assumed that it includes early detection of violators.

Along with the development of systems for recognizing individuals and monitoring human behavior, the fears in the society arise on the possibility of total surveillance of people by the state or the use of information obtained during surveillance without knowledge and consent of people for selfish purposes. An article reflecting these concerns has been published, for example, in *The New York Times* (Schneier 2020), and a copy is presented on its author's Web site.[30] The article mentions a number of advanced and sophisticated recognition technologies. People can be identified at a distance by their heartbeat or by their gait. Modern cameras can read fingerprints and patterns of the iris from a distance of several meters. In addition, it

[30] https://www.schneier.com/essays/archives/2020/01/were_banning_facial_.html.

is possible to recognize a person by mobile phone number, car number, credit card numbers, etc.

Currently, the legal regulation in this area cannot be considered satisfactory.

2.5.3 Prevention of Victims

Approaches to another manifestation of human factors in malicious actions—the "potential victim"—take into account the possibility to manage people who unwittingly perform this role and are aimed at ensuring their survival and preventing fatalities. There are methods for analyzing such situations and managing humans at the mass or individual level. The mass-level methods are in many ways similar to the approaches to the public in emergency situations of natural origin or technology-induced (see Sect. 2.3). An analytical overview of terrorist attacks of various kinds and risk factors affecting the public and rescuers is presented in the article collection Schweitzer and Fox (2009). In order to study collective behavior in these conditions and develop measures of special services to rescue people, simulations based on continuous, discrete or combined models described in Sect. 2.3.2 are used.

The individual-level approach is implemented in the recommendations on behavior under a threat and in conditions of malicious actions. Thus, the National Anti-Terrorist Committee of the Russian Federation has published the *General recommendations to citizens on behavior in the event of a threat of a terrorist act*.[31] These recommendations include provisions concerning:

- maintaining emergency preparedness;
- handling suspicious items "that may turn out to be explosive devices";
- evacuation;
- behavior in a crowd;
- behavior when taking as a hostage, including that in airplane;
- behavior at the threat of a terrorist act.

These provisions combine psychological aspects (assessment of the situation, personal behavior, interaction with terrorists, etc.) and objective characteristics of the situation (location, shelters, exits, time intervals, etc.).

On the website of the Federal Security Service of the Russian Federation (the Russian abbreviation is FSB), in the section "Tips of professionals", three documents are posted that contain recommendations for citizens in connection with threats of terrorist acts.[32] One of them coincides in content with the above *General recommendations...* The other was prepared as a result of a meeting of representatives of the FSB's Special Purpose Center with Moscow schoolchildren and teachers in 2004 and is entitled: *How to behave when being kidnapped and becoming a hostage of*

---

[31] http://nac.gov.ru/rekomendacii-po-pravilam-lichnoy-bezopasnosti/obshchie-rekomendacii-grazhdanam.html.

[32] http://www.fsb.ru/fsb/supplement/advice.htm.

terrorists. The recommendations presented in it complement the content of the first document, especially with regard to the behavior of hostages at critical moments. The third document refers to 2006 and concerns the events preceding the acts of sabotage and terrorism. It indicates typical signs of preparation for such acts that can be visible to the public, the stages of preparation, the set of methods used by terrorists, and the specifics of their living in apartments.

Despite the obvious benefits of such recommendations, they also have a negative side, noted, for example, in the publication of the French newspaper *Le Monde* on April 29, 2009[33]: when becoming known to potential terrorists, they contribute to changing the strategy and tactics of their actions.

[33] Presentation in Russian: https://www.inopressa.ru/article/30Apr2009/lemonde/kgb.html.

Chapter 3
Approaches to Human Factors from the Standpoint of Designer

3.1 The Role of Designer

3.1.1 Specificity of the Role

In the previous chapter, the approaches to human factors were presented depending on the human role in interacting with a technical system. All four categories of the roles considered in the context of this interaction—personnel, the public, vehicle occupants in crashes and people under conditions of malicious actions—manifest themselves during the operation of a technical system. The presented methods of accounting for human factors and managing them are applicable to various operational processes (including maintenance) and possible events in the operation stage (among which there are emergencies).

However, human factors also act before the start of operation and even before the existence of a technical system in material form—in the stage of its design. During this stage, they manifest themselves in the role of designer. This role, like the others discussed earlier, is complex one and falls into a number of components, or constituent "subroles". They are "played" by system developers, engineers of analytical or testing departments, managers and others involved in the creation of a system. A generic characteristic inherent in the complex role of designer and all its components can be considered the interaction of a human with a technical system that exists in the form of an immaterial (ideal) object—a project.

The designer's task is to create an image of the future system (intended to meet certain social needs), ensuring the effectiveness of its functioning and compliance with safety requirements during operation. Based on this task, designer must take into account influence of human factors on characteristics of the system under development that will work in nuclear power industry, air, space or ground transportation, civil infrastructure or other areas. We consider this influence differentially, so far using four categories of the human roles. However, human factors are also reflected in the design process itself, affecting its results and thereby determining an additional anthropogenic contribution to the consumer and operational qualities, reliability and

Y. Spirochkin, *Human Factors and Design*,
https://doi.org/10.1007/978-981-19-8832-5_3

safety of the system being created. All this makes it necessary to take into account various manifestations of human nature embodied in the role of designer and manage them when designing.

3.1.2 The Conception of Design

Design is the first of the five main stages in the life cycle inherent in technical systems of all types, as shown in Fig. 3.1. The presented diagram demonstrates that the approaches to human factors described in Chap. 2 are applied predominantly in the operation stage of a technical system (within the red frame). At least, some of these approaches are also applicable in the manufacture of system elements, in their supply and construction of the technical system as a whole, as well as in its decommissioning (included in the yellow frame). Such expansion of the area of applicability is possible due to the existence of the system as a material object in all post-design stages—at least at the level of elements—and the physical nature of human interaction with it.

Immaterial form of a system in the design stage determines specifics in the manifestations of human factors in this stage and, accordingly, approaches to them. This chapter has three tasks:

(1) to reveal features of the designer's role;
(2) evaluate the adequacy and completeness of the existing approaches to human factors embodied in other roles from the designer's standpoint, including management methods;
(3) identify ways to manage the design activities themselves in order to effectively use the positive and compensate for the negative aspects of human nature in the process of creating new technical systems.

Before proceeding to these tasks, let us supplement the content of the term *design* defined in Sect. 1.3, in linguistic and technical aspects. For example, the corresponding noun in Russian is "proektirovanie". In German, two different nouns are used: "Projektierung" in mechanical engineering and "Plannung" if they are talking about construction. The Russian term and the first German one originate

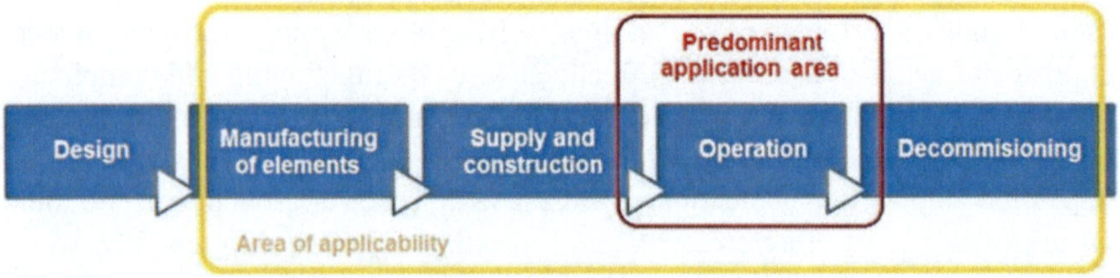

Fig. 3.1 Main stages of the life cycle of a technical system and the areas of approaches to human factors

from the Latin word "projectus" that means literally "thrown forward". The *Big Soviet Encyclopedia* interprets it as "the process of creating a project—a prototype or an archetype of a supposed or possible object or state". The modern definition is contained in the Russian national standard (GOST R 57193-2016) which reproduces the provisions of the relevant international documents on standardization. According to one of them—(ISO/IEC/IEEE 24765:2017)—the first meaning of design is as follows: "[process] to define the architecture, system elements, interfaces, and other characteristics of a system or system element". Another meaning is "phase of development concerned with determining what documentation will be provided in a product and the nature of the documentation". The result of design as the process (the phase of development) is information that "provides the detailed implementation-level physical structure, behavior, temporal relationships, and other attributes of system elements".

However, these standard definitions seem to be too general and incomplete. The author's definition of the term in question, presented in Sect. 1.3, from the point of view of designer—the developer of a technical system—can be expressed as follows. *Design* is a set of types and stages of work on preparation, based on specified requirements, of established technical documentation aimed at ensuring the post-design stages of the life cycle of the system being developed—from the manufacture of its elements to decommissioning inclusive. This definition reflects the important characteristics of the design:

(1) complex composition of design activities (they cover many types of work);
(2) dividing the design process into stages;
(3) design basis or data source (requirements for the system being developed);
(4) content and result of the design work (preparation of technical documentation, the list and format of which are established by standards and (or) terms of the design contract);
(5) purpose of the design (ensuring the subsequent stages of the life cycle of the system being developed).

Some comments should be given. The technical systems discussed in this book consist of many elements functioning on various physical principles and manufactured or built using various technologies. For example, aircraft includes airframe structure, one or more propulsion systems, power supply system, control system, navigation system, etc. The elements of nuclear power plant are nuclear reactor, equipment components and pipelines, buildings and structures, safety systems, control systems and others. The development of such heterogeneous elements and their integration into a complex requires specialized activities in a number of areas: mechanical engineering, construction, schematic and electrical design, physical research and structural analyses, mock-up, testing, etc. These types of design activities are usually implemented by different organizations.

The design process can be considered as solving a special type inverse problem: based on the requirements that determine the necessary behavior of the developed technical system in operation, the designer must determine its constructive image: in the terminology of (ISO/IEC/IEEE 24765:2017)—architecture, system elements, etc. At the lowest level, the constructive image implies geometry characteristics (shape, dimensions) of structural parts of the system and mechanical properties of materials. In this context, a direct problem is calculating a response of the technical system with known parameters to specified factors during operation (operational conditions, applied loads, maintenance or repair processes and so on). For structurally complex technical systems, designed for long-term operation under extreme conditions, solving even direct problems (problems of mathematical physics), despite the use of modern hardware and software, can be a difficult task. The difficulty of solving inverse problems in general, and more so, the problems on determining the structure and parameters of the system that should ensure its functioning under specified operational conditions with acceptable reaction levels (as well as meeting other established requirements), is immeasurably higher. Mathematically, such inverse problems are characterized not only by computational, but also by methodological complexity, including fuzziness and the absence of unambiguous solutions.

The problem of complexity is versatile; it also has a fundamental aspect. We will pay attention to it in Sect. 3.1.4; here, we consider the methodological side. Methodological complexity can be managed, first of all, by dividing the design process into stages corresponding to different levels of determining the constructive image of the system being developed, starting with the rough and generalized and moving toward the precise and detailed.

Thus, in the Russian Federation, the interstate standard (GOST 2.103-2013) establishes the following stages for the industrial design process: development of technical proposal, preliminary design, technical design, development of working design documentation.

In the first of these stages, possible solutions that meet the specified requirements are found, substantiated and compared, including solutions based on existing prototypes (GOST 2.103-2013). The content of the work performed by designers during this stage is defined by the standard (GOST 2.118-2013). The required result is the establishment of operating principle and placement of the system components that perform the main functions. The level of detail "should be sufficient for comparative assessment of the variants under consideration". In the stage of development of technical proposal, the search area for a design solution is narrowed so that, if not to exclude, then to minimize its ambiguity as much as possible. Technical proposal is drawn up in the form of documents provided for by the standards of the Unified System for Design Documentation.[1] After adoption and approval, it becomes the basis for preliminary or technical design.

The subject of the second stage—preliminary design—is development of technical documents "which should contain fundamental design solutions that give a general idea of the purpose, structure, operating principle and overall dimensions of the product being developed, as well as data determining its main parameters" (GOST 2.103-2013). Requirements for implementation of preliminary design are regulated by the standard (GOST 2.119-2013). In this stage, designers can consider variants for

[1] The Russian abbreviation of this system is ESKD ("Edinaya Sistema Konstruktorskoi Dokumentatsii").

the system under development and its components. The established general constructive image, despite the uncertainty in the details, allows designers to carry out approximate calculations to confirm serviceability, solve the main issues of manufacturing technology, organize operation support and evaluate the level of safety and some other characteristics of the system. As a result, the ambiguity of the design solution is already covered (ideally) by the range of variable values, in which parametric optimization is possible. Depending on the achieved depth of elaboration and a number of other factors, for example, the industrial or experimental purpose of the system being created, preliminary design serves as the basis (after adoption and approval) for the development of technical project or immediately working design documentation.

The stage of technical design is aimed at determining "the final technical solutions that give a complete picture of the product being developed, and the input data for the development of working design documentation" (GOST 2.103-2013). Implementation of the work in this stage is governed by the standard (GOST 2.120-2013). The remaining ambiguity is reduced, in fact, to the choice of optimal variants for certain components based on results of mock-ups' or prototypes' tests. When creating experimental technical systems and furthermore unique ones, for example, launch vehicle rockets or space ships, the stage of technical design may be absent. If this stage exists, then the design materials prepared during it, after adoption and approval, are used to develop working design documentation.

In the last design stage, a set of working design documentation is being prepared. The content of the work to be performed is defined by the standard (GOST 2.103-2013) only in a general form; it may differ in different branches of technology. The work may include the production and testing of one or more prototypes, the correction, if necessary, of design documentation, etc. The working design documentation, the types and completeness of which are established by the standard (GOST 2.102-2013), is intended for companies that manufacture elements of a technical system or built it. This documentation also includes operational documents according to the standard (GOST R 2.601-2019), which are delivered to the consumer (operator) of the system. Operational documents contain information necessary for the use of the system for its intended purpose, its maintenance and repairs, storage and transportation of its elements, their disposal, etc.

Significance of the five characteristics proposed within the framework the above definition of "design" is determined by the fact that neglecting any of them is fraught with negative consequences—ranging from the pure quality of the developed product to its complete failure.

Outside of Russia, the design stages are not established in regulatory documents, but in practice, the process is always iterative and includes a number of steps or phases. The main of them are[2] (Pahl et al. 2007; SEBoK 2021):

- task clarification (problem definition);
- conceptual design;
- embodiment design (preliminary design);

[2] Their names used in different publications vary.

- detail design.

Sometimes, only, the last three stages are distinguished (Fielding 1999; Anderson 2010).

The design methodology can be found in the fundamental manuals given above, as well as in an earlier publication translated into Russian (Jones 1982).

Design methods aimed at finding the best solution that meets a number of well-defined criteria are covered by the concept "optimal design". It involves the construction of an objective function that takes into account these criteria and the search for the values of the system parameters corresponding to the maximum or minimum of this function (depending on the formulation of the optimization problem). When determining optimization criteria, certain indicators are used that characterize performance (for example, required reliability or durability), technical perfection (minimum weight or material consumption, required level of safety), economic efficiency (cost of creation, cost of ownership, cost recovery), etc. Optimal design achieves its goal to the greatest extent when using computer software tools such as computer-aided design (CAD) and computer-aided engineering (CAE).

Regardless of the degree of optimization, design is currently carried out almost exclusively on computers. Basically, it can be considered as an activity related to the development and use of information and mathematical models describing the properties of the technical system being created (geometrical, mechanical, etc.), as well as the processes of its life cycle (manufacturing, assembly or construction, operation, diagnostics, maintenance, etc.).

A characteristic feature of design as a stage of the life cycle, at which the creation of new technical systems inheriting certain features of the systems of the previous generation is carried out, is the availability of feedback in the form of taking into account the experience of using these prototypes. Such feedback is also implemented by correcting errors identified during investigations of accidents and catastrophes, including those caused by human factors. The importance of this feedback is noted, for example, in the preface to the Russian edition of the book Salvendy (1987)[3] where it is stated (with reference to American ergonomics specialists) that if the study of human factors during the operation of a technical system "does not contain data of interest for design, its value is small or (more categorically) zero".

Along with increasing safety, the presence of an information flow of from operation to design, which ensures the correction of unsuccessful design decisions (in cybernetics terms, this is negative feedback), gives the entire industrial and social infrastructure associated with the creation and operation of technical systems the stability necessary for its existence.

There is another side to design. Like any complex and multi-stage human activity carried out by large teams over a long period of time, it requires organization and management. This side can be called "design logistics".

---

[3] The Russian edition of the book was issued in 1991–1992 by the Publishing House *Mir*, Moscow. The preface to it was written by V. P. Zinchenko and V. M. Munipov, editors of the Russian translation.

Based on the presented characteristics of design activity, we will consider the main roles of people involved in it or the components of the integral role of "designer".

3.1.3 Components of the Role

The division of the role of designer into components is due to the need to perform a variety of specialized engineering works related to the development of a technical system, as well as the necessity for their planning, coordination, control and the implementation of other management functions. The components of the role correspond to the directions of these works and management functions; they may be different in different branches of technology.

For example, in Russian rocket and space industry, representatives of the following engineering professions participate in the design process[4]: planners, analysis engineers, drafting and design engineers (actual designers), technologists, testers, etc. Planners are engaged in determining the overall constructive image of the system being developed, distributing specialized tasks between performers (divisions of the main project organization or other organizations), supporting cooperation, presenting and defending projects and so on. They make initial design decisions based on existing prototypes, generalized dependencies between the required functional and design parameters, experience, other available data, formal or informal knowledge. In subsequent iterations, the results of decision assessments by various analysis methods and based on various criteria are taken into account. These assessments are performed by analysis engineers in the fields of ballistics, aero- and gas dynamics, structural dynamics and strength, control, reliability, etc. Analyses, or calculations, are carried out using mathematical (numerical) models that describe the properties of the system being created with an accuracy corresponding (ideally) to the depth of its structural elaboration achieved, the expected conditions of its operation and functioning processes. Calculations that take into account the development of processes and events over time, i.e., dynamic analyses, are performed in the form of simulations. Drafting and design engineers transform the generalized constructive solutions substantiated by such analyses into assembly units and structural parts and formalize the results in the form of working design documentation. Technologists determine the possibilities of implementing system elements in terms of structural materials and production technologies. Testers check design solutions on mock-ups and experimental samples. The functions of design logistics are assigned to general and chief designers, heads of work directions and specialized (thematic) departments.

When creating aircraft, the role of planner is not allocated; the corresponding functions are covered by the complex role of designer. The other roles listed above are retained taking into account the characteristics inherent in aircraft. In particular, instead of ballistics, controlled flight in the atmosphere is analyzed; considerable

[4] On the example of NPO *Energia*.

attention is paid to aerodynamic design, avionics[5] development and ensuring fatigue strength of the structure under long-term cyclic loads. The logistics of design is similar to logistics in the rocket and space industry.

When developing Russian nuclear power plants, the design functions are shared among several types of organizations:

- institutes that carry out scientific leadership of projects in the field of physics and safety of nuclear technologies (National Research Center *Kurchatov Institute*, State Scientific Center of the Russian Federation—A. I. Leipunsky Institute for Physics and Power Engineering);
- design bureaus developing reactor installations (JSC *OKB Gidropress*, JSC *OKBM Afrikantov*);
- planner institutes that design NPP in the whole, perform the general layout of thermal and mechanical or other equipment components, and coordinate all construction work (for example, JSC Atomproekt);
- companies that develop the main equipment components (main circulation pumps, steam generators, turbine units, electromechanical devices, etc.).[6]

Within these functions, the following constituent subroles of the role of designer are performed: physics engineers (responsible for neutron-physical and thermos-hydraulic design of reactor cores), drafting and design engineers (developing thermal and mechanical or other equipment), analysis engineers (substantiating performance, strength and reliability of equipment components), materials scientists, designers of building structures, safety systems, control systems, etc. The functions design logistics are assigned to the general designer, the chief designer (person or design organization), the planner (person or planner organization), the main organization for materials science, the chief engineer of the project, heads of work directions and heads of departments.

The list of subroles that makes up the composite role of designer can be expanded when considering other industries. But, from all the diversity, both presented and left unmentioned, it is already possible to distinguish three constituent roles that characterize the design activity in the minimum necessary way. These are design engineer (who creates a constructive image of the system being developed), analysis engineer (who substantiates its performance using mathematical modeling and simulation) and head of design works (responsible for the logistics of design). The selected constituent roles allow us to determine:

- standpoint of a "generalized designer" when assessing how the existing approaches to human factors meet design basis and design methodology;

[5] A complex of electronic on-board systems for communication, navigation, display, flight control, collision prevention, meteorological observation, registration of flight parameters. In military aircraft, such systems also include weapons control systems, including radars, acoustic systems (sonars), electron-optical systems for detecting targets, electronic warfare support, etc.

[6] Among organizations of Russian nuclear industry, JSC NIKIET stands out—this institute combines functions of reactor installation design and scientific leadership in projects.

- areas where attention to human factors, manifested in the design process itself, should be focused;
- appropriate opportunities for design activity management.

3.1.4 Issues of Design Activities

There are fundamental problems inherent in design activity that can affect significantly on work processes and lead to design errors. Such problems include uncertainty, complexity and limited resources.

Uncertainty is usually divided into

- uncertainty of nature;
- uncertainty of adversary and
- uncertainty of goals.

Uncertainty is reflected in one way or another in design basis of the system being created and, accordingly, in design results. According to academician N. N. Moiseev,[7] "it is impossible to overcome this uncertainty by formal methods …—it is in the nature of things"; it is only possible to narrow the degree of uncertainty by known mathematical methods (Moiseev 1979). When taking into account uncertainty, probabilistic methods are usually used. They are based on the approximation of representative statistical data by appropriate probabilistic model (probability distribution function and parameters) and include analytical inference methods. Statistical data on a natural phenomenon may be incomplete, and because of this, the resulting probabilistic model may have limited application. Such shortcoming of the probabilistic model describing the uncertainty of nature has led to fatal error in determining the maximum height of tsunami wave, which caused the accident at the *Fukushima Daiichi* NPP (see Sect. 1.2). There may also be errors in the approximation of statistical data.

Similar shortcomings of the probabilistic methods manifest themselves when assessing the uncertainty of adversary. In order to improve probabilistic assessment, Bayesian approach[8] is applied. This approach is based on the possibility of updating an a priori evaluation of the probability of an event (which is interpreted initially as an amount of epistemic confidence, or the strength of belief, rather than a frequency and may not reflect any statistics), using newly obtained statistical data on such events. In the book Silver (2012), the corresponding procedure is described on the example of predicting malicious actions (however post factum)—the terrorist attack on the World Trade Center in New York (this example is reproduced in Sect. 4.2.2).

---

[7] Moiseev, Nikita Nikolaevich (1917–2000), is a Soviet and Russian scientist who made contributions to mechanics, numerical methods in the theory of optimal control, the theory of hierarchical systems, simulation methods, design automation and a number of other scientific and technical fields.

[8] It is named so after its author, the Reverend Thomas Bayes (1701–1761), who proposed the theorem (Bayes' theorem) on which this approach is built.

It should be noted, that updating a priori forecasts for such events is possible only when new relevant data are received, and this data can only appear as a result of an extraordinary event that has already occurred. Thus, the price for accuracy may be unacceptably high.

The uncertainty of goals can be caused by the presence of several of them (which creates the problem of choice), the lack of clear ideas about the objective function of a project, etc. An example is the Baltic NPP project,[9] in the objective function of which the emphasis was placed on economic indicators (payback), but its estimates were not consistent with the real market conditions, and as a result, the project was stopped.

Paradoxically, sometimes, the uncertainty of goals can also play a positive role, opening up new opportunities. In the late 1940s–early 1950s, work was being carried out in the Soviet Union to create a thermonuclear weapon (H-bomb) and a means of delivering it to the territory of a potential enemy—an intercontinental ballistic rocket. The main input data for the design of the rocket were the mass of the payload delivered along a ballistic trajectory. At the beginning of the development of the rocket, this mass was unknown because the H-bomb and the rocket were being created simultaneously. The uncertainty has been eliminated at the level of the country's top leadership, which demanded to develop the rocket for a bomb with the mass up to 10 tons.

However, the mass of the H-bomb after the completion of its design turned out to be much smaller, and this made the R-7 rocket (see Fig. 3.2), created to early 1957 by S. P. Korolyov,[10] excessively powerful for military purposes. These purposes were better met by the projects developed around the same time by other chief designers. The inconvenience of maintaining a rocket using liquid fuel (kerosene) and a cryogenic oxidizer (oxygen) in constant combat readiness also played a role. But, the creation of the R-7 rocket which became little in demand in militarily terms allowed our country, the USSR, to be the first to launch a human into space.

Complexity is an equally serious problem in the design process. The multifaceted nature of this problem and its manifestations in structural, methodological and computational aspects were mentioned above in Sect. 3.1.2. Now, we consider complexity as a fundamental characteristic. It is inherent in dynamic systems consisting of large number of elements with insufficiently traceable and controlled connections. The lack of traceability and controllability implies both incomplete knowledge about the entire variety of connections between elements and the inability to determine with a given accuracy the parameters of existing connections under any influences on the elements.

The following systems can be considered as complex ones:

(1) designed objects that include many structural and functional elements, the properties and interfaces of which correspond to new principles of functioning

[9] A nuclear power plant under construction in Kaliningrad region, Russia.

[10] Korolyov, Sergei Pavlovich (1907–1966), is one of the main creators of Soviet rocket and space systems, Chairman of the Council of Chief Designers in the USSR.

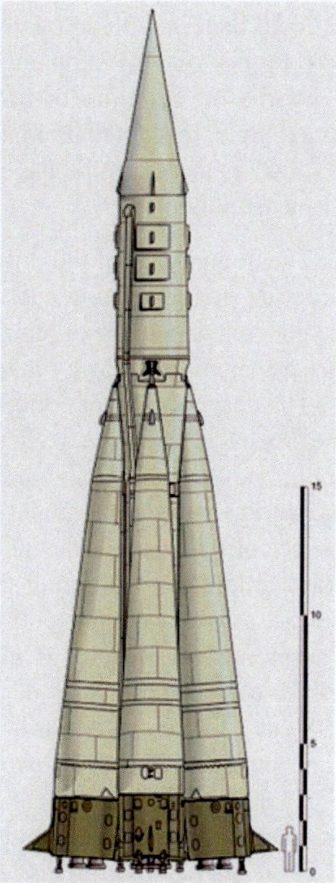

Fig. 3.2 First version of the R-7 rocket, tested in 1957. Reproduced from https://ru.wikipedia.org/wiki/P-7. CC BY-SA 3.0

(significantly different from prototypes), in particular when these elements themselves are under development;

(2) mathematical models—especially those that are used to solve nonlinear dynamic problems and have a large number of degrees of freedom;
(3) design teams consisting of representatives of various organizations and departments that perform work functions and interact with each other at different levels with differences in the understanding and implementation of these functions.

The complexity is largely due to the uncertainty discussed above. In the systems that belong to the first or second of these types, the causes of uncertainty and, accordingly, complexity are as follows:

- incomplete or unreliable knowledge about the physical processes associated with the new principles of functioning;
- non-linearity, due to which the coupling parameters determined for one observed interval of interaction of elements turn out to be incorrect when transferred to any other interval;
- finite resolution when representing variables in digital electronic devices used for numerical modeling and simulation, such as processors, as a result of which

disturbances with an intensity lower than the existing resolution are not detected (but they may develop uncontrollably in the future);
- stochastic variation of properties, especially due to hidden imperfections, for example, initial technological defects or aging, which does not allow developers or users to determine the exact properties of each product sample without direct measurements.

The complexity of the third type of systems (design teams) is determined by the structure of cooperation and the organization of design activities at different levels (in general, within the organization, within the department), as well as psychological aspects (including those related to the fields of industrial or social psychology).

The consequence of complexity is an accumulation of uncontrolled deviations in the process of solving a design task, which ultimately leads to an unpredictable result, sometimes very far from the expected one. This result may be an error in the layout or interface, non-compliance with the specified requirements, incorrect estimates of the performance of the designed system, an increase in the timing and cost of design. An example demonstrating the influence of complexity on the timing and cost of design, as well as on the entire project as a whole, is the International Experimental Thermonuclear Reactor. The ITER project has started in the mid of 1980s. The author of the book was repeatedly involved in the design work in the period from 1989 to 2007, so he is familiar with a number of problems firsthand. A significant contribution to the complexity of this project (the general view of ITER is shown in Fig. 3.3) is being made not only by the novelty of the principles of functioning based on thermonuclear fusion and the corresponding scientific and technical problems, but also by the design logistics associated with the work of the international central team and the participation of national teams distributed across several countries. The prolonged duration of the design work has led to a shift in the planned completion date of construction to 2025 (when the total period of the creation of ITER will be about 40 years!) and to an increase in its initial cost by almost 4 times (Ipatova 2019).

Big projects require resources corresponding to their scale in time, money, material, technological aspects, as well as relevant human resources, including intellectual ones. Limited resources can lead to poor quality and even to the collapse of a project. Resource provision refers mainly to logistics, but limited technological resources can manifest themselves in aspects that are difficult to manage, e.g., those related to the level of development of science and technology in the country. An example of such manifestation was the Soviet "lunar" project, in which the existing limitations concerned primarily the engines of the N-1 rocket. In the 1960s and the first half of the 1970s, the USSR did not have the technological capabilities to create high-thrust engines similar to the main engines of the American *Saturn* V rocket (the thrust of each of its five oxygen-kerosene F-1 engines was about 6.8 MN at ground level; the total thrust was more than 34 MN).

Under these circumstances, it was decided to equip the first stage of the N-1 rocket with well-developed (to that time) and reliable oxygen-kerosene NK-15 engines. The relatively low thrust of each engine (about 1.510 MN at ground level) was compensated by their number–30; the total thrust ranged from 45 to 50 MN depending

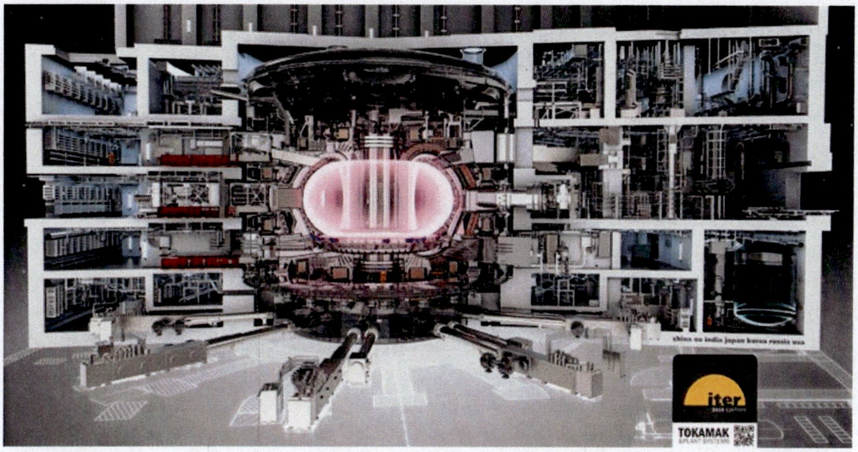

Fig. 3.3 General view of ITER. Image by Oak Ridge National Laboratory. Reproduced from https://www.flickr.com/photos/37940997@N05/41783636452. CC BY 2.0

on modifications. The placement of NK-15 engines on the N-1 first stage rocket is presented in Fig. 3.4a. Figure 3.4b shows for comparison the engines of the first stage of *Saturn* V.

A large quantity of NK-15 engines (a kind of record for the number of engines on one rocket in the history of cosmonautics) led to negative consequences, including the appearance of a turning moment when even one engine failed, the need to shut down the diametrically opposite one and as a result, loss of thrust. These effects were

a b

Fig. 3.4 Engines of the first stage of: **a** The Soviet N-1 rocket (Reproduced from http://www.aerospaceweb.org/question/spacecraft/q0196.shtml) and **b** The American *Saturn* V rocket (*Photo* Bernt Rostad. Reproduced from https://www.flickr.com/photos/brostad/3553239979. CC BY 2.0) discovered during flight tests. All four test launches of the N-1 rocket ended in failure at the first stage of the flight. The approach itself—detecting problems during test flights and not on ground stands—was the result of another limitation: due to lack of time, money and material resources, stands for dynamic and firing tests of the first stage or the entire rocket assembly were not built. In May 1974, work on the N-1 project was discontinued.

It is obvious that in all the design problems discussed above, human factors have manifested themselves in one way or another. When solving such problems, the abilities of designers to manage uncertainty and complexity, as well as compensate for limited resources, can be crucial. These professional abilities would undoubtedly contribute to improving the quality of designs, efficiency and safety of the technical systems being created.

3.2 Evaluation of Existing Approaches to Human Factors

3.2.1 The Standpoint of Designer

In order to evaluate existing approaches to human factors, which were overviewed in Chap. 2, from the standpoint of designer, we will define this standpoint taking into account the characteristics of the design presented in Sect. 3.1.2, the components of the role of designer considered in Sect. 3.1.3 and the issues of design activities discussed in Sect. 3.1.4.

Our evaluation will be based on the awareness that design tasks necessarily include ensuring safety during human interaction with technical systems. In the process of interaction, these systems can become a threat due to intentional or unintentional actions of humans who created them or who use them in the operation stage. In this regard, human factors can manifest themselves as a source of threat, and at the same time, they can be embodied in potential victims of threatening interaction. On the other hand, a good way out of some dangerous situations becomes possible (at least at the current stage of technology development) only due to the appropriate actions of human as a control link of a human–machine system.

Thus, the dependence of the level of safety on human factors is extremely complex: some aspects of these factors determine the negative or positive control effects on technical systems and through them on humans themselves, while others are related to the conditions for preserving human health or life when responding to such effects. Such ambiguity of human interaction with technical system should be taken into account in the design.

From all this, we determine the standpoint of designer as follows:

- approaches to human factors that are not directly related to design are excluded from the evaluation area;
- of interest are data on human interaction with a technical system that may affect its structure and functions in connection with ensuring safety;

- to form design basis of the system, a phenomenological description of this interaction using physical, including mechanical parameters (spatial coordinates, time, intensity of forces, etc.), is necessary;
- this description should correspond to the established design technology, which is influenced by uncertainty, complexity and limited resources;
- any safety characteristics used in the design should be expressed in quantitative, measurable parameters.

When evaluating, we will not take into account the national origin of the technology: in the context of globalization, it does not seem so significant. For example, as reviewed in Chap. 2, the regulations used in Russian aviation, wheeled vehicles and rail transport reproduce relevant rules and standards adopted in the United States, the EU or at the UN level. Russian rocket and space technology, as well as Russian nuclear power industry, were developing independently and stand out against this background. However, international cooperation in space—both on the delivery of commercial payloads of various countries to Earth orbit using Russian launch vehicles, and on the ISS project—has led to the actual harmonization of Russian and foreign standards for the design of space systems. As for Russian nuclear power industry, its regulatory framework and technical solutions are based on the same principles, criteria and methods that are used all over the world, and nuclear power plants built according to Russian projects are characterized by safety levels comparable to the corresponding data for nuclear facilities in the USA, in Europe and other countries (Spirochkin 2019).

3.2.2 Coverage of Human–Machine Interaction and Systematization of Approaches

Approaches to human factors discussed in Chap. 2 cover all the roles performed by humans in the stage of operation of a technical system. With a certain degree of generalization, based on the materiality of the system elements and the physical nature of human interaction with them (through production machines, construction or installation equipment, and other tools), these approaches extend to other post-design stages of the life cycle of the system, presented in Fig. 3.1.

Taking into account human factors and managing them form a set of data, methods of their processing and measures related to ensuring safety in the entire spectrum of human–machine interaction, based on the biological, psychological and social characteristics inherent in humans. An attempt to systematize approaches to human factors from the standpoint of designer allows us to distinguish the following components in this set:

- regulation of safety conditions (in the form of requirements, rules, recommendations, etc.);
- design and construction solutions (which must meet regulatory provisions);

- information and mathematical models (used to substantiate the solutions being developed);
- experimental data (test results for the purpose of studying features of human–machine interaction, checking design and construction solutions, confirming their compliance with established requirements (certification), verifying and correcting models, as well as statistical information, including that in relation to emergency situations).

Below, in Sects. 3.2.3 to 3.2.6, the listed components are evaluated in their current state. Sect. 3.2.7 summarizes the results of evaluation; in addition, the applicability of existing approaches to human factors to the global trend of tightening safety requirements is analyzed.

Organizational measures are not explicitly included in the above scheme. They are presented in Sect. 2.2.5; design as an area of human activity and the results of this activity are undoubtedly related to them. Some aspects of this relationship—in terms of the effect of the safety culture of operating personnel on the design basis and the interdependence of rescue procedures and characteristics of rescue equipment used in emergency situations—are evaluated below, in Sect. 3.2.3. But, in general, organizational measures and methods of managing human factors are considered separately, in Sect. 3.3, in relation to the management of design activities.

3.2.3 Regulation

As follows from Chap. 2, the analysis of human factors and their management in nuclear power industry, in the field of transport and in civil infrastructure are subject of regulation at various levels. The upper level of regulation corresponds to legal documents, which are mandatory for use. In Russia, such documents include federal laws, technical regulations and codes introduced by them, government resolutions, federal regulations and rules being in force in the industries important for safety. A lower status is typical for standardization documents: interstate and national standards, standards of organizations, etc. According to the Federal Law *On Technical Regulation*,[11] their application is voluntary, but in the industries, most sensitive to safety, e.g., in nuclear power industry, compliance with certain standards is in practice mandatory (Spirochkin 2019).

In the Russian Federation, international conventions and codes are effective (for example, on the rescue of people in distress at sea), as well as guidelines of international organizations in which this State is a member, for example, ICAO or IAEA. The status of international conventions and codes is the highest; guidelines are at the level of recommendations, but, as a rule, they are recognized by Russian regulatory authorities and are observed by designers. Over time, their provisions are incorporated in national documents of different levels.

[11] The Federal Law No. 184-FZ of December 27, 2002, as amended on July 2, 2021.

The regulatory function is also inherent in technical specifications and technical requirements stipulated in contracts between customers and contractors. These documents define design basis of the technical system to be developed. They include the requirements of the rules and standards that must be met, as well as the features of consumer, operational and other qualities that the system must possess. Implementation of these provisions is mandatory within the application area established and limited by the terms of contracts.

For the first category of human roles—personnel—the ergonomic design of workplaces and the corresponding support of work processes are regulated (see Sect. 2.2.2), rescue in emergency situations that may arise during operation (Sect. 2.2.3), formation and support of safety culture (Sect. 2.2.5).

The ergonomic requirements expressed in quantitative terms directly affect the structural design of the created technical system. The relevant provisions concerning the equipment of the aircraft crew decks are well-developed; they are established in the airworthiness standards and aviation rules, which relate to the upper level of regulation. Obviously, this degree of development is due to the high cost of intervention in the work of pilots from the point of view of safety, especially during the rapid processes accompanying the flight. Some regulatory provisions (for example, concerning the design of the control panel and informing operators) are borrowed from aviation and space technology by other industries.

Requirements for the ergonomic support of work processes in the manufacture of elements of technical systems, the building or assembly of systems as a whole, their maintenance, repair and decommissioning are also important for determining the structural design. They are mainly contained in standardization documents, including standards of organizations. The most general requirements of this kind are established in interstate and national standards. At this level, for example, activities in the field of quality management are regulated by the standards (GOST ISO 9000-2011), (GOST ISO 9001-2011).

Aspects of accounting and management of human factors related to the rescue of personnel in emergency situations at sea are regulated at all levels. From the standpoint of designer, the establishment of detailed requirements for sea vessels and offshore facilities to ensure evacuation, including equipping them with rescue equipment, is very rational, and the requirements for the design parameters of this equipment are also very useful. No high-level regulatory documents were found concerning the rescue of personnel operating other technical systems, but of course, rescue procedures and appropriate equipment exist. Obviously, they are determined by industry documents (including technical specifications) and technical requirements for individual developments.

The formation and support of safety culture is provided for in the FNP-level regulatory documents in force in Russian nuclear power industry, in the IAEA safety standards and the ICAO guidelines. The relationship between the quality of personnel management and the level of safety culture in the operating organization, on the one hand, and the requirements for the design of a nuclear power plant or aircraft, on the other, is not obvious. However, depending on the quality of management and safety culture, it is possible that the operating personnel either does or does not certain effects on the controlled technical system, including situations when such effects are necessary for the implementation of work processes or correction of emerging deviations. Therefore, the variability of personnel actions, as well as the possibility of inaction (as an extreme case), should be provided for in the design basis, if this may affect safety.

In nuclear power industry, the conditionality of accidents and catastrophes by shortcomings in personnel management and safety culture (identified during investigations of emergency events) is taken into account, and this accounting sometimes goes beyond the design basis. For example, the lessons learnt from the accident at the second power unit of the *Three Mile Island* NPP, which was caused by the inability of operators to detect a coolant leak in the primary circuit timely and their erroneous intervention in the automatic operation of the reactor emergency cooling system (shutdown of pumps), led to very large-scale measures, including:

- directives issued by NRC for all nuclear power plants operating in the United States on safety improvement;
- organization of constant supervision, systematic assessment of the level of operation, maintenance and engineering support of nuclear power plants;
- development of detailed evacuation plans for the population around a nuclear power plant in case of an emergency;
- changes in the structure of NRC itself.

The cause of the disaster at the unit 4 of the Chernobyl NPP was recognized as a combination of the actions of personnel during the testing of the RBMK-1000 reactor in a dangerous mode with its design flaws.

The International Nuclear Safety Advisory Group (INSAG), established under the auspices of the IAEA, identified, among other things, the following (INSAG 1993):

- general insufficiency of safety culture in nuclear matters, both at the national and local levels;
- lack of necessary information amid the personnel and understanding of the NPP features important for safety;
- errors of the personnel that led to the violation of existing instructions and testing program.

The Chernobyl disaster, like the *Three Mile Island* accident, dealt a serious blow to the global nuclear power industry as a whole. From 1986 to 2002, not a single new nuclear power plant was built either in North America or in Western Europe (due to the anti-atomic social movement and the decline in the profitability of nuclear power generation in the face of increased safety and security measures). The social and technical consequences in the USSR were also significant. The impact of the Chernobyl disaster on the development of the Soviet and Russian nuclear power industry expressed in the suspension of the construction and design of nuclear power plants for more than a decade and in the rejection of the direction of RBMK. Changes were made in the regulation concerning safety in the design of nuclear power plants with other types of reactors. For example, the following new provisions appeared (NP-001-15):

- reliability analysis when substantiating safety of NPP "should be carried out taking into account … errors of personnel";
- safety control systems "should be designed in such a way as to prevent the possibility of disconnection … by the operating personnel within 10–30 min after their automatic start, but not to prevent the correct actions of operator in an accident".

As for the comparison of nuclear power industry and civil aviation in this direction of ensuring safety, it should be noted that neither American *Airworthiness Standards* (CFR Part 25) nor Russian *Aviation Rules* (AP-25 2009) contain such requirements. Visual indications of the aircraft configuration and the position of the controls, audible alarms and warnings implemented in accordance with these documents can be ignored by the flight crew. Automatic flight control systems (autopilot, autothrottle control, altitude control, etc.) can be disabled by the pilot for one reason or another. As a result, it is possible to overcome the established safety barriers, as demonstrated by the Swiss cheese model (see Sect. 2.2.4).

Regulation of approaches to human factors in the role of the public, which is considered as a "target" of possible dangerous impacts and "passive" object of management, seems to be adequate to the extent while safety requirements are expressed in quantitative parameters: allowable margins of impacts, spatial and temporal characteristics of the conditions for evacuation of people under threats, etc. These requirements and parameters are established in the relevant regulatory documents including standards (listed in Sect. 2.3.1). Organizational measures and procedures on the protection and rescue of the population in emergency situations of various origins, technical means for rescue and protection are also regulated. With regard to the rescue of the public in distress at sea, the same regulation applies that discussed above with respect to personnel of sea vessels and offshore facilities.

The approach to the public as a multitude of active subjects (agents) that can be controlled (see Sect. 2.3.2), partially alternative to its consideration only as a "target" or "passive" object of management, is practically not reflected in the available regulatory documents.

The interaction of vehicle occupants with vehicle structure during a crash is a very developed area of regulation, in which quantitative parameters of hazard factors, injury and survival criteria are established: acceleration values, deformations of the life volume, injury indices, evacuation time, etc. (see Sects. 2.3.1 and 2.4). The procedures for determining such parameters and their allowable values used in air and ground transport are included in relevant rules, standards and manuals. Because of the high speed of crash processes, which makes it problematic to have any active control directly during the impact, passive safety requirements are mainly imposed on the vehicle.

Taking into account human factors in the role of attacker is expressed, first of all, in protecting technical systems from sabotage, terroristic or other unauthorized malicious actions. This is ensured by systems of physical protection of nuclear facilities, aviation security services, technical means of pre-flight inspection, as well as measures for structural protection of flight crew compartments from violent intrusion (see Sect. 2.5.1). The development and operation of such safety barriers are regulated at the highest level: in Russia—by the requirements of FNP, the *Air Code* and *Aviation Rules*, in the United States—by the provisions of the relevant sections of CFR. Some requirements and provisions are directly applicable as part of the design basis. There are also regulatory documents with the status of recommendations.

Along with the formation of safety barriers at nuclear facilities against malicious actions from the outside, protection against possible insiders is regulated. The relevant provisions are given, for example, in CFR and relate to monitoring the trustworthiness and reliability of individuals who have access to information about sources of potential danger and (or) to the sources themselves. Similar rules (but not of an open nature) are applied in Russian nuclear power industry. The trustworthiness and reliability of the personnel are monitored to some extent in aviation and other industries important for safety, but this is not related to design requirements. Judging by the published open materials, the situation when a member of personnel who has dangerous intentions or intruder from the outside, having overcome the established barriers, gets access to the control of a technical system is beyond regulation.

Early detection of violators using profiling or technical devices is carried out in the Russian Federation in accordance with industry documents that do not have the status of regulations, codes or rules; there is only one standard that regulates only terminology (see Sect. 2.5.2).

As for the role of potential victim of malicious actions, taking into account human factors and managing them are aimed at saving hostages without fatal consequences. When designing vehicles, buildings and infrastructure facilities, in some cases, it is possible to use the provisions of regulatory documents on safety in emergency situations (see Sect. 2.3.1), which make it possible to assess the movement of human masses, ensure rapid evacuation or predict dynamic effects. With regard to individuals taken hostage, there are recommendations on behavior aimed at reducing the risk of harm from intruders (see Sect. 2.5.3). Some of these recommendations may contribute to the creation of technical systems that ensure, on the one hand, the protection and rescue of hostages and, on the other hand, the neutralization of attackers.

* * *

Finally, it is useful to express our evaluation of the state of regulation in a quantitative form (thereby expressing the consistency of the designer's methodology). To do this, we consider the following indicators:

- the level of regulation, R, with the basic values equal to: 1 (if there are mandatory documents), 0.5 (if there are only documents of non-mandatory (voluntary) status or limited area of application) or 0 (in the absence of any regulatory provisions);
- the nature of the data used in regulation, D: 1 (quantitative data), 0.5 (there are provisions for determining quantitative data), 0 (no data available);
- compliance with the design technology, T: 1 (full compliance), 0.5 (partial compliance), 0 (no compliance).

In order to take into account the existing incompleteness of regulation, unresolved problems, heterogeneity of evaluations for different industries, etc., we will use a reduction coefficient for the specified basic values less than 1. The listed indicators and the values assigned to them reflect the subjective point of view of the author. The evaluation results, or ratings, are presented in Table 3.1; they can be considered simply as an example.

Table 3.1 Regulation as a component of the approaches to human factors: evaluation results (ratings)

Role: personnel

Indicator	Evaluations in relation to			Average rating of the indicator
	Ergonomics	Rescue in emergency situations	Safety culture	
R	0.8	0.7	0.6	0.7
D	0.7	0.6	0.5	0.6
T	1	1	0.5	0.83

Average rating for all indicators: 0.71

Role: the public

Indicator	Evaluations in relation to		Average rating of the indicator
	The public as a passive object	The public as a multitude of active subjects	
R	1	0	0.5
D	0.8	0	0.4
T	0.8	0	0.4

Average rating for all indicators: 0.43

Role: vehicle occupants in crashes

Indicator	Evaluations in relation to			Average rating of the indicator
	Aircraft	Wheeled vehicles	Rail vehicles	
R	1	1	1	1
D	0.9	0.9	0.9	0.9
T	0.9	0.9	0.9	0.9

Average rating for all indicators: 0.93

Role: people under conditions of malicious actions

Indicator	Evaluations in relation to			Average rating of the indicator
	Protection against malicious actions	Early detection of violators	Potential victims	
R	0.8	0.4	0.4	0.53
D	0.7	0.5	0.4	0.53
T	0.7	0.4	0.4	0.5

Average rating for all indicators: 0.52

Despite the fact that these results are conditional and inaccurate, they demonstrate differences in the depth of regulation for different roles and show areas for which additional development is required. To the greatest extent, the consideration of human factors is regulated in relation to the design of vehicles according to the crash safety criteria (the average rating for all indicators is 0.93), i.e., in relation to the role "vehicle occupants in crashes". The design requirements are least developed taking into account the active behavior of the public (the average rating for this role is 0.43). The ratings of regulation in regard of the remaining roles are within these boundaries. Areas of regulation characterized by lower ratings need additional rulemaking.

3.2.4 Design and Construction Solutions

We will consider design and construction solutions—the second component of the approaches to human factors noted in Sect. 3.2.2—as data on the composition and layout of the technical system, its structure, materials, control and safety devices, etc. These data are determined in the design process taking into account the participation of human factors in the life cycle of the system. In Chap. 2, a part of such solutions is presented, reflecting a large number of existing ones. We focus on some important characteristics of this totality, first of all, on the relationship with other components.

The requirement that design and construction solutions comply with established regulatory provisions shows the imperative of the latter for design. However, there is also a reverse effect. In fact, regulatory codes, rules and standards are developed based on the experience of designing and operating technical systems. Design and construction solutions embodied in successful advanced samples of machines and facilities become—with a certain generalization—the content of regulatory documents, acquiring the form of requirements and (or) recommendations for design of such technical systems in the future. As a result, the provisions of regulatory codes, rules and standards (primarily expressed in quantitative parameters) implicitly depend on the structural characteristics on the basis of which they are formed. Applying them to other design and construction solutions, including non-traditional, can be problematic. This feature of regulation is noted in Sects. 2.4.3 and 2.4.4.

Regulatory documents, like prototypes, help to manage uncertainty and complexity when creating technical systems and help to minimize design errors, but their conservatism toward changes sometimes significantly limits the possibilities for improving design through new principles of construction or the use of new materials. Regulatory provisions are subject to revision (or updating) only periodically. For example, for the ASME Boiler & Pressure Vessel Code[12] containing rules for design of NPP equipment and pipelines, this period is 3 years (ASME 2019). Updating the Consolidated List of standardization documents used as mandatory

[12] ASME Boiler & Pressure Vessel Code is a standard developed by The American Society of Mechanical Engineers (ASME) and recognized as an international regulatory document.

in the State Corporation Rosatom is carried out as needed, but at least once every 5 years.[13]

If the properties of the developed technical system differ significantly from those provided for by regulatory provisions, additional (a kind of adaptive) methods are used to ensure safe operation and confirm compliance of new properties with the established requirements. So, in the case of the ASME Boiler & Pressure Vessel Code, taking into account the development of technologies, additions and changes are made to the relevant sections of this document with their annual publication. In order to establish, when urgently needed, rules for the design of structures or the use of materials not covered by the existing provisions of the code, so-called "Code Cases" are formulated and published (ASME 2019). The Russian Federal Service for Environmental, Technological and Nuclear Supervision (Rostechnadzor) permits design organizations to substantiate design and construction solutions in nuclear power industry that do not fall under the scope of existing FNP documents by submitting documentation containing analysis methods and results or experimental data for their detailed assessment by experts. According to the FAA rules, for aircraft of unconventional design, it is possible to substantiate its properties that provide the occupant protection in a crash by comparing with existing prototypes (see Sect. 2.4.4).

The relationship of design and construction solutions with models, numerical analyses (calculations) and experimental data will be analyzed below, in Sects. 3.2.5 and 3.2.6. In the meantime, we will turn to such an important aspect of approaches to human factors (in the field of design and construction solutions) as protection used in technical systems against its negative manifestations.

There are several directions of such protection. The first of them can be considered protection against unintentional misuse—the so-called "fool protection". This is implemented, for example, by forming the controls of the technical system in such a way as to help the operator avoid mistakes caused by ignorance, negligence, inattention, forgetfulness, fatigue (see Sect. 2.2.2). In the working processes, the following elements of the "fool protection" are provided:

- signals from sensors, the absence of which, due to improper behavior of the operator, blocks further use of the system;
- control of the sequence of operations performed by mechanical or electronic devices, as a result of which it is impossible to miss any actions or disrupt their sequence, etc.

The second direction of protection under consideration, which minimizes human errors, is the use of automatic control systems (also mentioned in Sect. 2.2.2), up to the complete exclusion of the human operator from the control loop. This direction reaches its highest level in unmanned vehicles operating in space, air, on water or under water, in automatic cars and trains, in fully robotic production facilities. Along

[13] According to the regulation on standardization in relation to products (works, services) for which requirements associated with safety in the field of nuclear energy use are established, as well as processes and other standardization objects related to such products/Approved by the Decree of the Government of the Russian Federation No. 669 of July 12, 2016.

with the undoubted advantages of automation, being partial, it can contribute to a decrease in the qualifications of personnel. For example, it is difficult for pilots of aircraft with automatic control to acquire the necessary flight skills; in addition, there is a risk of losing previously acquired ones because they are not used.

If in a critical situation there is a need for manual control, the available qualifications may not be enough. Thus, the unprofessional actions of the crew with the autopilot disabled (in combination with other circumstances) have led to the crash of the Boeing 737-500 aircraft of *Tatarstan Airlines* (flight TAK 363) during landing at Kazan airport on November 17, 2013 (MAK 2013).

On the other hand, the operation of completely unmanned vehicles at the current level of technology development can be considered safe only outside the human habitat.

An example of a corresponding threat is the crash of the *Orion* UAV in the Ryazan region, Russia, on November 16, 2019. It happened at the distance 70 m from the residential buildings of *Listvyanka* village, and only, by a lucky chance, there were no victims (Afonsky and Felipyonok 2019).

The third direction of design and construction solutions for protection against negative manifestations of human factors is protection taking into account deliberate malicious actions. This direction includes, for example, the development of the structure of the cockpit door (with fasteners) that can resist violent intrusion from the outside (see Sect. 2.5.1). The strength of the door will not save the situation when a crewmember becomes the attacker (two such cases out of several known are mentioned in Sect. 2.1).

A possible way to protect against dangerous, unauthorized actions of any person sitting in the pilot's seat would be to timely recognize and block them. The implementation of this method requires the use of special "smart" control tools as part of a technical system.

Design and engineering solutions should ensure the safe operation of a technical system in the specified conditions of its use. However, human interaction with the system during operation requires another important property of these solutions: in addition to safe operation, they must ensure the preservation of human lives in an emergency situation, preventing it from escalating into a catastrophe. In the initial phase of an accident, this can be implemented due to proper control of a vehicle or facility by its personnel and (or) due to the functioning of automatic control systems. This possibility is considered in Sect. 2.4.3 regarding emergency landing of aircraft. Examples of automatic systems designed for emergency management during the beginning and initial development of an accident are the safety systems of nuclear power plants: protective, providing and controlling. Some information about them can be found in the previous book of the author (Spirochkin 2019).

If it is impossible to prevent the accident development, people can be saved by removing them from the technical system that is doomed to destruction. Relevant design and construction solutions include ejection seats and escape capsules for military aircraft crews, emergency rescue systems for spacecraft crews and other rescue equipment (see Sect. 2.2.3). To ensure the survival of vehicle occupants in a crash (in the final phase of the accident development accompanied by destruction), the properties of safe destructibility (crashworthiness) are used, as well as safety systems: safety belts and inflatable pillows, and after crash—equipment for evacuation, localization of the release of dangerous substances, notification of rescuers, etc. (see Sect. 2.4). For unmanned aerial vehicles, if it is impossible to cope with an emergency situation that has arisen in flight, preventive self-destruction is used to avoid possible catastrophic consequences in the event of a fall in a populated area.

Ensuring the survival of vehicle occupants in a crash through the properties of crashworthiness requires a change in design principles. Traditionally, the structure of any technical system is developed using the principle of exclusion of limit states associated which destruction (due to plastic deformations, loss of stability, crack growth, etc.). The exclusion is achieved by fulfillment of the criteria of strength and integrity[14] of the structure in the following form (Spirochkin 2019):

$$C_k = (\mathcal{S}_k \geq \mathcal{F}_k)(k = 1, 2, \ldots K) \quad (3.1)$$

where $\mathcal{S}_k$ is a measure of strength and integrity under loads applied, taking into account the k-th limit state;

$\mathcal{F}_k$ is a measure of the loading that leads to this limit state;

K designates the number of limit states and, accordingly, the number of strength criteria.

The measures $\mathcal{S}_k$ and $\mathcal{F}_k$ depend on the nature of loading and potentially realizable limit states (which correspond to the equality of $\mathcal{S}_k$ and $\mathcal{F}_k$) and represent the values of forces, stresses, deformations, the number of loading cycles, fatigue damage, etc.

Crashworthiness (related to emergency situations) is associated with the consideration of a slightly different set of limit states and criteria. Some of them still relate to the strength and integrity of the structure, i.e., they serve to ensure safety by excluding destruction—at least those parts of the structure whose destruction is unacceptable. This part of the limit states and strength criteria is invariant with respect to the contact of the system with human. The other part arises only in connection with the human presence. So, in the case of a vehicle with occupants, accelerations during the crash impact, forced movements of occupants, structural parts and equipment components, the life volume inside deformable compartments, the time of evacuation after the accident, etc., are taken into account (see Sect. 2.4). The criteria used to minimize injuries and ensure survival differ from the criteria of strength and integrity; moreover, a programmable destruction of the structure is necessary for their implementation. The set of such criteria (it varies for different vehicles) includes indicators of injury to various parts of the human body (Sect. 2.4.6), energy absorption capacity and maximum deformation of energy-absorbing zones of the structure, special requirements for programmable destructible elements (Sect. 2.4.7), etc.

---

[14] Integrity (the term used alongside with strength) is a generalized characteristic of a structure as a set of components, assemblies and parts. In the case of pressure-retaining structures (vessels), integrity characterizes the continuity of the material and the tightness of the vessel, taking into account the possibility of cracks or loosening of the detachable joints. Cracks in the walls of vessels or loose joints may not affect the strength of the main structure (up to a certain limit), but leakage of liquid or gaseous contents through them indicates a violation of integrity.

Fig. 3.5 Sets of limit states and criteria used when ensuring safety of technical systems, taking into account human factors (set A relates only to the strength and integrity of the structure; set B relates only to the human tolerability of potentially dangerous effects)

The change in design principles, based on the safety requirements in emergency situations, affected not only vehicles. For example, in new nuclear power plants of the generation 3+, core melt catchers are used that belong to the category of localizing safety systems working in the final phase of a nuclear accident. In case of destruction of the reactor vessel, such a catcher filled with sacrificial material keeps the molten radioactive substances within the protective shell, preventing their release into the environment. Sacrificial material (of a certain composition and microstructure) enters into a physicochemical interaction with the radioactive melt, which leads to a decrease in its temperature and volumetric density of the energy release (Gusarov et al. 2005). The conditions providing acceptable levels of these parameters, as well as the time interval characterizing the resistance of the core melt catcher,[15] can be considered as criteria for its design.

The relationship between the sets of limit states and the criteria applied when ensuring the safety of a technical system taking into account human factors is illustrated in Fig. 3.5. Set A refers to the exclusion of structural failure, i.e., preservation of its strength and integrity. Set B corresponds to maintaining the levels of factors potentially threatening a human below the limits of tolerability. For normal operation conditions, these sets are independent. In an accident, they become interconnected. For example, it is possible in principle to ensure safety in a crash by increasing the strength of structural elements and (or) building up protective equipment. But, if this leads to a significant increase in the cost of the system, then such measures are unacceptable for economic reasons. In this case, a compromise must be implemented in the design of the system between strength and integrity, on the one hand, and the level of safety determined by the limits of the tolerability to threatening factors and the possibilities of evacuating people from the danger zone, on the other.

When developing a vehicle, increasing the strength of its structure in order to ensure safety may lead to exceeding the limits of the tolerability determined by impact accelerations. The area of intersection of the two sets, designated as A & B, characterizes the simultaneous consideration of the limit states of the structure and the limits of human tolerability to threatening factors in an emergency situation. In

[15] This time interval is not established in regulatory documents, but in practice, according to Dimitar Popov, the value 72 h is used.

this area, programmable destruction of structural elements aimed at absorbing the energy of these factors is permissible, but at the same time, the strength and integrity of at least the last protective barriers are ensured. Thanks to this strategy, the general conditions for human survival are preserved although certain injuries are possible due to the achievement of individual tolerance limits.

* * *

A quantitative evaluation of the state of design and construction solutions can be performed according to the methodology described above in Sect. 3.2.3. As indicators, we take the following:

- the level of prototypes, *P*, with the basic values equal to: 1 (if standardized solutions exist), 0.5 (if there are solutions acceptable as prototypes), 0 (if there are no standard solutions and prototypes);
- completeness of safety provision, *C*: 1 (sufficient), 0.5 (partial), 0 (none);
- applicability for new designs, *A*: 1 (full applicability), 0.5 (partial applicability), 0 (non-applicability).

The heterogeneity of evaluations for various branches of technology and the existence of unresolved problems are taken into account, as before, through a reduction coefficient of less than 1. The evaluation results (ratings that should be considered as an example) are shown in Table 3.2.

According to these results, design and construction solutions are the most developed in terms of ensuring the safety of personnel (the average rating for all indicators is 0.72). Along with the acceptable level of design of technical systems, that take into account the public as a passive object, no similar adequate solutions have been found for cases of its active behavior; therefore, the overall (average) rating is low (0.35). The relatively low values of the considered indicators for aircraft crashes (0.6) are due to the fact that not all possibilities for rescuing passengers and crews are realized in modern civil aircraft, for example, escape capsules are not used. The concept of crashworthiness is also not fully applied.

3.2.5 *Information and Mathematical Models*

The main purpose of using information and mathematical models when creating a technical system is to substantiate design and construction solutions. In the context of human factors, such models also serve to analyze and clarify planned organizational measures.

The information models presented in Chap. 2 reflect the logical structure of human–machine interaction and the conditionality of accidents and catastrophes in the general case (see Sect. 2.2.4), as well as the manifestation of various factors in automobile accidents (Sect. 2.4.6). Mathematical models describe processes of human interaction with a technical system and the environment using quantitative

Table 3.2 Design and construction solutions as a component of the approaches to human factors: evaluation results (ratings)

Role: personnel

Indicator	Evaluations in relation to			Average rating of the indicator
	Protection against negative manifestations of the human factor	Protection and rescue in emergency situations	Ensuring survival in serious accidents	
P	0.7	0.8	0.8	0.77
C	0.7	0.7	0.7	0.7
A	0.7	0.7	0.7	0.7

Average rating for all indicators: 0.72

Role: the public

Indicator	Evaluations in relation to		Average rating of the indicator
	The public as a passive object	The public as a multitude of active subjects	
P	0.7	0	0.35
C	0.7	0	0.35
A	0.7	0	0.35

Average rating for all indicators: 0.35

Role: vehicle occupants in crashes

Indicator	Evaluations in relation to			Average rating of the indicator
	Aircraft	Wheeled vehicles	Rail vehicles	
P	0.6	0.7	0.6	0.63
C	0.6	0.7	0.5	0.6
A	0.6	0.7	0.5	0.6

Average rating for all indicators: 0.61

Role: people under conditions of malicious actions

Indicator	Evaluations in relation to			Average rating of the indicator
	Protection against malicious actions	Early detection of violators	Potential victims	
P	0.7	0.5	0.4	0.53
C	0.7	0.5	0.4	0.53
A	0.7	0.5	0.4	0.53

Average rating for all indicators: 0.53

variables and parameters, with a certain degree of detail (accuracy). Modeling in one form (logical relations) or another (mathematical expressions) covers all human roles and all operational situations.

When analyzing the actions of personnel, the range of modeling is quite wide: along with the above-mentioned logic schemes, the behavior of a human operator is described using mathematical models of automatic control theory and reliability theory, based on empirical data on human tolerability to mechanical effects and probabilistic approximations of the recorded response characteristics (Sect. 2.2.1).

With regard to the public as a "passive" object of management (Sect. 2.3.1), modeling is not so developed, and a threshold approach prevails: comparing the levels of dangerous effects with established allowable limits and implementing regulated organizational measures when these limits are exceeded. In some cases, analytical models of fluid flow are used to describe the movements of a human crowd. The factors determining the active behavior of the public (Sect. 2.3.2) partly lie in the field of psychology. Quantitative analysis, which has experimental substantiation, is currently possible only at a low level of manifestations of conscious behavior, by considering analytical models of the flow of human "fluid" and taking into account the excited state of the crowd by varying the flow rate. To describe the human interaction in the crowd more accurate, molecular dynamics models are developed, but they are based on hard-to-verify assumptions, such as artificial forces of a sociopsychological nature.

Simulation models, which are computer-oriented mathematical models of processes and are increasingly used in studies of human mass behavior, including emergencies, with all the variety of possibilities for displaying individual and collective actions, have a number of shortcomings:

(1) due to their inherent algorithmic nature, they do not provide demonstrable dependencies of behavior parameters on acting factors (unlike, for example, the description by formulas);
(2) the prerequisites underlying them, the features of algorithms for «playing» the situations and other characteristics that determine the limits of applicability are, as a rule, hidden for users (this is typical for commercial software and is a potential source of modeling errors);
(3) hypotheses that allow analysts to build quantitative models of conscious behavior can be substantiated only in a statistical sense (they do not necessarily correspond to each specific situation being studied);
(4) to obtain even approximate results, well-defined input data and a long simulation time are required (both are usually unavailable in the initial stages of designing a technical system);
(5) the simulation results are limited—at least in the works known to the author—by spatial and (or) temporal parameters of behavior and do not contain information about the intensity of mechanical interaction of people with technical systems, including that in emergency situations (and such information is necessary for design).

Mathematical models being involved in analyses of vehicle crashes also belong to the category of simulation ones, but they are applied only to mechanical processes. These models are constructed using numerical methods: LPM, MEM or FEM (see Sects. 2.4.3 to 2.4.7). The latter method has, apparently, unlimited potential in accurately describing the human–machine interaction during the collapse of a structure, based only on the geometric characteristics and properties of materials. Due to this feature, it is also applicable for determining the parameters of simplified models corresponding to the first two methods (Sect. 2.4.6). Modern finite element models are usually very complex. Their complexity is caused by:

(1) the connected behavior of a technical system, a human and the environment in which emergency situation develops;
(2) mechanical diversity of the structure of the system and the human body, differences in the properties of their parts and components of the environment;
(3) non-linearity of behavior processes, when properties of the model components are not fixed, but depend on the parameters of impact and response to it;
(4) a large number of variables needed to describe these processes (degrees of freedom);
(5) sensitivity of the simulation results to errors in the specified properties and to the accuracy of the approximation of variables.

One of the indicators of complexity is the number of degrees of freedom of the built model or a comparable number of finite elements N. This indicator increases with a more detailed meshing of the analyzed system into finite elements (reducing their size), and the meshing process is associated with controlling the accuracy of modeling. Along with the increase of N, the time spent on obtaining a solution grows; therefore, the possibility of increasing accuracy is provided by using more powerful and sophisticated computer hardware and more efficient software codes, that implement, for example, parallel computing. However, the dependence of the accuracy (estimated by the modeling error) on the size of the elements is theoretically substantiated only for solutions of elliptic equations using polynomial shape functions (Frid 1972). For problems with energy dissipation, corresponding to non-linear dynamics of a structure under conditions of destruction, there is no such substantiation. Nevertheless, it is possible to demonstrate the effect of the number of elements on the modeling error for a finite element solution using the examples of vehicle crash simulations overviewed in Chap. 2. Data from some of these examples, including data on modeling errors, are generalized in Table 3.3.

The error ε characterizes the difference between the calculated parameter of the structural response of the vehicle $\mathcal{V}_c$ and the corresponding experimental value $\mathcal{V}_e$ (negative variable: "lack of accuracy"):

$$\varepsilon = -\left|\frac{\mathcal{V}^{\text{э}} - \mathcal{V}^{\text{р}}}{\mathcal{V}^{\text{э}}}\right| \times 100\%. \tag{3.2}$$

As the parameter of the structural response, unless otherwise specified, the maximum acceleration value $a_{\max}$ is considered. The values of velocity given for the examples 1 to 4 correspond to the vertical velocity of the aircraft at the impact.

In Fig. 3.6, the modeling (simulation) error is presented depending on the number of finite elements, i.e., in the form of a mathematical function $\varepsilon = \varepsilon(N)$. This function (curve) approximates the data in Table 3.3 related to the examples 2, 4, 5 and 6, selected because it is possible to evaluate the error for them based on the

Table 3.3 Data from examples of vehicle crash simulations carried out using mathematical models: generalization

Example	Time period	Number of finite elements N	Velocity, m/s	τ [a], c	$a_{\max}$, g	Error modulus $\|\varepsilon\|$, %
1. Crash landing of *Buran* (Spirochkin 1987)	1983–1987	~ 10^3	3	0.16	17	
2. CID/Boeing 720 (McComb et al. 1987; Fasanella et al. 1990)	1984–1990	~ 2×10^2	5	0.15	17	30
3. Emergency landing of Mi-34 (Lyakhovenko et al. 1989; Spirochkin and Shenk 1990)	1989	~ 10^2	10	0.075	40	20[b]
4. The Boeing 737 fuselage section at vertical drop test (Jackson and Fasanella 2001)	2000–2001	~ 1.4×10^4	10	0.1–0.15	14–22	25
5. High-speed train collision with a rigid obstacle (Tang et al. 2016)[c]	2016	~ 2.5×10^5	80	0.25		20[d]
6. Crash test of the Ford *Taurus* (Munyazikwiye et al. 2018)	2018	~ 10^6	56	0.12–0.13		4–20[e]

[a]Acceleration pulse duration (see Fig. 2.1)
[b]The simulation error of the maximum deformation of the cabin
[c]In the simulation, in addition to finite elements, mesh-free approximations were used
[d]Maximum simulation error for internal forces
[e]The lower value refers to the maximum deformation of the front part of the car, and the greater corresponds to ASI

results of experiments (or more detail calculation). For the last example, the average value of the lower and the greater estimates of the error is used.

The function in Fig. 3.6 looks like a hyperbolic one, similar to saturation curves in thermodynamics. This function shows a significant improving accuracy with an increase in the number of finite elements to a certain characteristic value, after which a further growth of N no longer leads to the corresponding decrease of the error $|\varepsilon|$ due to the lowering of the derivative $\frac{d\varepsilon}{dN}$. It can be expected that with an unlimited increase in N, the curve asymptotically approaches zero. We see that the modeling accuracy corresponding to the error of 20% is achieved with the number of finite elements in the range from 150,000 to 200,000. Models with a greater number of elements have a smaller error, but their use increases the calculation time. At $N \sim 10^6$, it is from one to two days (see Sects. 2.4.6 and 2.4.7).

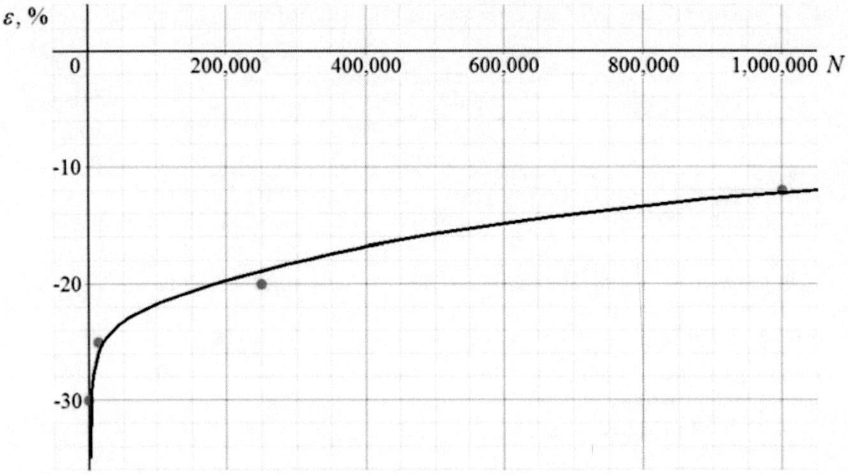

Fig. 3.6 Effect of the number of finite elements N on the modeling error ε in the simulations of vehicle crashes

Due to the long time to obtain a solution using detailed modeling, the need to have accurate data on the geometry and material properties of the analyzed system, numerical models are practically inapplicable when creating its generalized constructive image, i.e., in the initial stages of design (when there are no time resources or certainty in structural parameters). In fact, such models do not meet the minimum requirements for analysis tools, which are formulated as follows (Du Bois et al. 2000):

- accuracy of prediction of essential desired features;
- reasonable time to obtain results—no more than 12 h, regardless of model size, for use in design iterations and parametric studies;
- robustness—small variations in model parameters should not lead to large changes in model response;
- model development time shall take a reasonably short period, not exceeding 2 weeks.

The algorithmic, numerical nature of modeling by FEM, LPM or MEM does not provide an understanding of the dependence of safety indicators on structural parameters and operation factors. Obtaining such dependence is in principle possible by processing the results of computer simulation (dynamic analysis) in the field of these parameters and factors. To carry out such simulations, a numerical model based on input data with the necessary accuracy is needed. In addition, for a stable computational process, the system of dynamic equilibrium equations must have a certain condition number. Meanwhile, the accuracy of input data and the precision of calculations required for their stability have little in common with each other, and they usually do not correlate with the accuracy of the simulation results.

As follows from Table 3.3 and Fig. 3.6, even very detailed models containing about a million finite elements give results different from experimental data. The differences are most likely caused by fundamental reasons that practically do not depend on the degree of detail: errors in modeling the mechanisms of destruction and energy dissipation in a crash, as well as insufficiently accurate consideration of the contact interaction of densely located equipment components in the crushed interior space of the vehicle. Despite the reduction in the size of finite elements, in the FEM models presented in available publications, hidden technological defects are usually not taken into account, but it is their nature and localization that determine the features of the propagation of elastic waves in the structure and the accumulation of irreversible deformations, as well as the trajectories of destruction processes and their intensity levels.

In particular, such effects observed in practice as violation of integrity of welded structures due to avalanche-like opening of welds (see Sect. 2.4.7) are not modeled. In order to minimize the number of defects and compensate for their impact on crashworthiness, special design and technological measures are used, but the question of the predictability of such an effect in the design stage is still far from being resolved.

As for the numerical models of the human body used in crash simulations, their detail (demonstrated in Sect. 2.4.6) seems excessive for design purposes. Parameters of the dynamic response of body parts and internal organs, the mass of which is significantly less than the mass of the whole body and especially the mass of the vehicle, can be estimated without their direct modeling, for example, using shock acceleration spectra at the seat attachments points (Spirochkin 1993).

From the standpoint of designer, numerical modeling based on FEM, LPM or MEM is the most effective in the stage of detail design (see Sect. 3.1.2), when the use of complex models is adequate to the tasks of optimizing design solutions—in accordance with the purposes of optimization analysis, available input data and time resources. The area of optimization lies within the characteristics of the technical system found by simpler and more demonstrable methods. To find these characteristics in the initial stages of design, analytical functions (formulas), tabular data and graphical curves are used, approximating their dependence on operational factors and required safety indicators.

Theoretical, empirical and semi-empirical methods of this kind, which make it possible to determine the configuration, basic dimensions and properties of system components based on predicted operating conditions and established safety levels, are traditionally used in all branches of technology. For nuclear power facilities, they are contained, for example, in the relevant sections of the ASME Boiler & Pressure Vessel Code (ASME 2019) or in the corresponding Russian regulatory document (PNAE G-7-002-86). For airplanes and helicopters, the set of design dependencies reflects the provisions of American *Airworthiness Standards* (CFR Part 25) or Russian *Aviation Rules* (AP-25 2009); the dependencies themselves can be found in the design manuals (Fielding 1999; Anderson 2010; Asselin 2012; Torenbeek 2013). When designing vehicle rockets and spacecraft, relations similar to those given in the book Wijker (2008) are used, etc. According to the established terminology, presented, for example, in the European standard (EN 13445-1:2009), these methods belong to the category *design by formulas*.

To ensure safety taking into account human factors (in accordance with the criteria of human survival in accidents), another subset of dependencies is added, reflecting the relationship between the characteristics of the system and survival indicators (in accordance with the situation illustrated in Fig. 3.5). Such dependencies are used, for example, in the design of aircraft based on the concept of crashworthiness (USAAVSCOM TR 89-D-22C 1989).

Approaches using mathematical models and numerical calculation methods, primarily FEM, represent an alternative category: *design by analysis* (EN 13445-1:2009).

Human factors in the role of attacker are described by information models of the "violators' model" category, and these models may contain biometric data. To analyze the behavior of a violator, mathematical models and computer simulation tools are applicable, in particular discrete-event modeling based on Markov chains (see Sects. 2.5.1 and 2.5.2). When considering the role of a "potential victim" (in order to ensure his or her survival and prevent fatality), appropriate simulation models developed for the role category of the public may be useful (Sect. 2.3.2). These models can be built using assumptions about the "passivity" of the public (simplified continuum modeling) or its "activity" (discrete modeling, including agent-based one). The shortcomings of models of this kind, obvious from the standpoint of designer, were discussed above.

* * *

As before (see Sects. 3.2.3 and 3.2.4), we try to express our evaluation of the state of information and mathematical models (their compliance with the existing design technology) in a quantitative form. For this purpose, we select the following indicators:

- the nature of the data (variables and parameters) processed during the modeling and simulation, D, with the basic values equal to: 1 (if all data are countable, i.e., can be expressed quantitatively), 0.6 (if the data correspond to a threshold approach, to simplified, generalized approximation or are biometric ones), 0.4 (if the data are logical); 0 (if there no data are available);
- applicability in the design process, A: 1 (in any stage of design), 0.6 (only in the initial stages of design), 0.4 (only in the stage of detail design); 0 (in case of inapplicability);
- functionality, F: 1 (to substantiate design and construction solutions), 0.5 (for reference purposes), 0 (if there is no function).

The evaluation results (ratings) are presented (again just as an example) in Table 3.4.

As can be seen, higher ratings of compliance with the existing design technology (a kind of degree of perfection) belong to models that are used in simulations of vehicle crashes taking into account human factors (0.83) or simulations of personnel

Table 3.4 Information and mathematical models as a component of the approaches to human factors: evaluation results (ratings)

Role: personnel			
Indicator	Evaluations in relation to		Average rating of the indicator
	Information models	Mathematical models	
D	0.4	1	0.7
A	1	1	1
F	0.5	1	0.75
Average rating for all indicators: 0.82			

Role: the public			
Indicator	Evaluations in relation to		Average rating of the indicator
	The public as a passive object	The public as a multitude of active subjects	
D	0.6	1	0.8
A	0.6	0.4	0.5
F	1	0.5	0.75
Average rating for all indicators: 0.68			

Role: vehicle occupants in crashes			
Indicator	Evaluations in relation to		Average rating of the indicator
	The oretical, empirical and semi-empirical models	Numerical models	
D	1	1	1
A	0.6	0.4	0.5
F	1	1	1
Average rating for all indicators: 0.83			

Role: people under conditions of malicious actions				
Indicator	Evaluations in relation to			Average rating of the indicator
	Protection against malicious actions	Early detection of violators	Potential victims	
D	0.4	0.6	0.6	0.53
A	0.6	0.6	0.6	0.6
F	1	0.5	0.5	0.67
Average rating for all indicators: 0.6				

work (0.82). The least adequate models are those describing the behavior of people under conditions of malicious actions (0.6).

3.2.6 Experimental Data

The sources of experimental data that can be useful in the design of technical systems taking into account human factors are test results and statistical information obtained during the operation of existing systems, including emergency situations.

There are two main categories of tests accompanying the design: research and certification ones. Research tests are aimed at studying the human–machine interaction, checking (verification) of design and constructive solutions, substantiation and identification of mathematical models. These tests can be carried out using physical, for example, small-scale models of the developed system, simplified or partial mock-ups, operating or decommissioned prototypes of the system—with the main goal of solving design problems related to safety (see Sects. 2.2.1, 2.3.2, 2.4.2, and 2.4.4 to 2.4.7).

The purpose of certification (qualification) tests is to confirm the compliance of the created technical system with the established requirements. Such tests are carried out on existing samples of the system and, as a rule, do not pursue any research goals. During certification tests, inconsistencies may be identified that need to be eliminated in order to obtain permission to operate the system. Nevertheless, these tests, along with research ones, as well as statistical data on the operation of technical systems can provide valuable information about human—machine interaction, suitable for taking into account human factors in the design.

From emergency situations—accidents and catastrophes, data can be extracted on their scenarios, the behavior of the structure, the mechanisms and consequences of its destruction, the efficiency of energy-absorbing elements and the conditions in which people survive—in terms of life volume inside the vehicle, human body movements, acceleration levels, injuries, evacuation measures (see Sects. 2.3.2, and 2.4.2 to 2.4.7).

The importance of experimental data for obtaining knowledge about human–machine interaction, substantiation of design and construction solutions and mathematical models determines their indispensability in the whole. At the same time, an experimental test is usually an expensive undertaking, not to mention the price that has to be "paid" in an emergency situation. The high cost of full-scale experiments, together with an unacceptable degree of danger when reproducing accidents in real technical systems that can develop into disasters, leads to the need to replace such tests with small-scale physical tests or mathematical modeling and computer simulation.

According to results of a study conducted in connection with tests on aircraft and car crashworthiness, reducing the scale of modeling to 1:3 leads to a halving of the cost of testing, and with a ratio of 1:8, only, a quarter of the costs required for a full-scale experiment are required (USAAVSCOM TR 89-D-22C 1989). However, when preparing small-scale tests, technological difficulties arise in the manufacture of model elements that imitate thin-walled parts of a real structure, in particular, shells. In addition, it is impossible to simultaneously ensure similarity in structural dynamics and the loss of the bearing capacity of structural elements.

Table 3.5 Experimental data as a component of the approaches to human factors: evaluation results (ratings)

Indicator	Evaluations in relation to			Average rating of the indicator
	Research tests	Certification (qualification) tests	Statistics on operation	
A	1	0.5	1	0.83
F	1	0.5	0.5	0.67
Average rating for all indicators: 0.75				

Mathematical models are free from these shortcomings, but their substantiation can become a problem. To identify the model parameters in the conditions of nonlinear dynamic behavior, sophisticated experimental and theoretical tools are needed.

* * *

A quantitative evaluation of the state of experimental data from the standpoint of designer can be carried out (as before, as an example) using the following indicators:

- applicability of experimental data in the design process, *A*: 1 (full applicability), 0.5 (partial applicability); 0 (non-applicability);
- functionality, *F*: 1 (wide functionality: for verification of design and construction solutions, substantiation and identification of mathematical models, etc.), 0.5 (limited functionality: for reference purposes); 0 (if there is no function).

These indicators are considered invariant to the categories of roles in which human factors manifest themselves. The results of evaluation (as simplified as possible for example purposes) are shown in Table 3.5.

3.2.7 Generalized Evaluation, also Taking into Account More Tighten Safety Requirements

As our evaluations show (see Sects. 3.2.2 to 3.2.6), the existing approaches to human factors in the design of technical systems cover all human roles in interacting with systems and all operation situations. The considered components of these approaches: regulation, design and construction solutions, information and mathematical models, and experimental data are interrelated. Their condition reflects the current level of development of science and technology. In Tables 3.1, 3.2, 3.4 and 3.5, these components were evaluated quantitatively, as differentiated by the human roles and constituent indicators of each component. Subjective nature of the evaluations has allowed us to consider them so far only as examples. Now, we will try to move further: we will determine average ratings in the coordinates "components-roles", and then, we will obtain a generalized evaluation of the approaches to human factors. Perhaps

Table 3.6 Average ratings and generalized evaluation of the approaches to human factors

Role	Component of the approaches				Average rating for the role
	Regulation	Design and construction solutions	Information and mathematical models	Experimental data	
Personnel	0.71	0.72	0.82	0.75	0.75
The public	0.43	0.35	0.68	0.75	0.55
Vehicle occupants in crashes	0.93	0.61	0.83	0.75	0.78
People under conditions of malicious actions	0.52	0.53	0.6	0.75	0.6
Average rating for the component	0.65	0.55	0.73	0.75	
Generalized evaluation					0.67

some patterns will be discovered along the way, as well as problems and the potential to solve them. The results of finding average ratings and obtaining a generalized evaluation are presented in Table 3.6. It should be recalled that the highest rating on our scale is 1.

As can be seen from Table 3.6, the higher average ratings relate to the approaches applied to the role "vehicle occupants in crashes" (0.78) and the role "personnel" (0.75). The least developed are approaches to "the public" (0.55), which can be explained by the insufficient level of requirements for the design of technical systems as well as the relatively low level of design and construction solutions, taking into account active human behavior. The average rating of approaches to the role "people under conditions of malicious actions" is slightly higher (0.6); however, it is relatively low—obviously, due to a regulation deficiency or shortage of adequate design and construction solutions that ensure safety.

If we turn to comparing the average ratings for the components of the approaches, we can note a more "advanced" state of experimental data (the average rating is 0.75) or information and mathematical models (0.73). Regulation (with the average rating 0.65) or design and construction solutions (0.55) seem less developed: there are gaps in these components related to taking into account the behavior of the public as a multitude of active subjects (agents).

The generalized evaluation (0.67) characterizes the generally satisfactory state of the existing approaches to human factors in the design of technical systems. At the same time, the obtained value indicates the incompleteness of these approaches: figuratively speaking, they correspond to the "ideal" design technology ensuring safety only by 2/3. The "shortage" of ensuring safety "through design" (1/3) is compensated by organizational measures in the stage of operation, including the rescue of people in emergency situations, which cannot be excluded. Planning such measures contributes to maintaining safety at a level acceptable to society, but it is obvious that their implementation (especially in the form of rescue operations) is connected with an additional risk, the magnitude of which, including due to the influence of human factors, is difficult to predict. In practice, a compromise is implemented between safety requirements and the possibilities of meeting them, and the possibilities are determined not only by the state of science and technology, but also by economic factors.

The tightening of safety requirements imposed on technical systems in connection with resonant accidents and catastrophes of recent times causes a "shift" of the compromise toward reducing risk and, accordingly, more advanced ways of ensuring safety. The obtained evaluations show that the existing approaches to human factors have the potential for improvement, since none of the components of these approaches currently reaches the highest value: 1.

It is possible to improve all components and in relation to all roles, but it is necessary to focus efforts primarily where the ratings are lowest: regulation, as well as design and construction solutions to ensure safety of the public, in particular taking into account its active behavior (ratings 0.43 and 0.35). The second direction may be to pay more attention to regulation concerning human factors embodied in the "intruder" and the development of technical systems whose characteristics could increase the chances of potential victims to survive (current ratings are 0.52 and 0.53). In Sect. 3.2.3, the lack of regulation has been already mentioned which could exclude the possibility an attacker (including one from the personnel) controlling the system, as well as insufficient level of technical means capable of recognizing and blocking dangerous actions. To substantiate the design characteristics that could increase the chances of survival, it is necessary to develop information and mathematical models describing the interaction of people with the system in the conditions of malicious actions (the rating of their current state is 0.6).

The relatively low rating of design and construction solutions that ensure safety of occupants in vehicle crashes (0.61) is due to the fact noted in Sect. 3.2.4: not all possibilities for rescuing passengers and crews in emergency situations are currently being implemented. For example, if an accident during the flight of a civil aircraft occurs, control is carried out mainly by the pilot actions, and if it was not possible to stop the negative development of the situation, passive safety means are used. However, the concept of crashworthiness, which is the most complete embodiment of passive safety, is not fully put into practice. The current level of science and technology makes it possible to increase the chances of survival due to the active safety means used to rescue crews in military aviation and cosmonautics at critical moments of accidents (ejection seats, escape capsules, launch escape systems (LES or SAS). But, such design and construction solutions are not adopted in civil aviation—apparently, due to the significant cost of their development and implementation. Ensuring safety, taking into account possible accidents, should not depend only on the skill of pilots—a shift in the design paradigm[16] is necessary. This shift should include an increase of safety characteristics in the object function of the project compared to economic factors and a change in the ratio of compensating actions performed by the crew and automatic controls.

Another possible direction for improving approaches to human factors in order to increase safety when interacting with technical systems is the development of mathematical models of the public behavior. The current rating (0.68) reflects the insufficient compliance of the models describing its active behavior with the design technology; the shortcomings of the simulation models are listed in Sect. 3.2.5.

Along with the general, more than satisfactory state of the approaches to human factors in the role of personnel (rating 0.75), there are components of these approaches that are rated lower: regulation (0.71) and design and construction solutions (0.72). In Sect. 3.2.3, the problems that determine such a decline in ratings were noted:

- insufficient level of accounting for the variability of actions or the possibility of inaction of personnel in regulatory documents and the design bases of technical systems, despite the connection of these aspects with safety;
- the possibility of an emergency when a crewmember disables one or more automatic aircraft control systems that perform the functions of safety barriers.

In order to solve these problems, it is necessary to shift the design paradigm in the appropriate directions.

Thus, the generalization of quantitative estimates, which initially looked like very subjective, and the mutual comparison of the ratings obtained, allowed us to identify patterns inherent in current approaches to human factors, confirm the existence of problems (detected at the level of logic) and indicate the potential for increasing safety by solving them. A well-known shortcoming of the method for ranking and weighing of qualitative characteristics is that the results of their quantitative assessment based on the commonly used pairwise comparison may differ from the results of simultaneous comparison of a larger number of characteristics (Jones 1986). In our evaluation procedure presented above, an increased number of characteristics (indicators) participated. This gives us reason to hope that the possible errors are not so great as to fundamentally distort the results. In order to reduce the errors inherent in discrete expert estimates, an analytical procedure can be applied to reconcile them—for example, based on the determination of an eigenvector (Saaty 1980).

Some possibilities for increasing safety of technical systems, including a shift in the design paradigm, are discussed in Chap. 4, but first, we will turn to design activity management (according to the last task of this chapter, noted in Sect. 3.1.2). Such management is necessary to solve a number of problems: applied, related to a specific project, methodological, related to changes in design approaches and fundamental, which were discussed in Sect. 3.1.4. In design activity management, we will limit

[16] The design paradigm is understood as a set of concepts concerning the properties that a technical systems should have, principles of their development and supporting the subsequent stages of the life cycle, as well as the methods and models used; this set is generally accepted among specialists involved in the design.

ourselves to the "human dimension": aspects of human factors inherent in the role of designer and ignore other dimensions: financial, technological, informational, etc.

3.3 Design Activity Management in the Human Dimension

3.3.1 Levels of Management

The design processes should be provided with the necessary resources: scientific, regulatory, financial, human and others. When accumulating and using these resources, human factors manifest themselves, and they can significantly affect the course and results of the design. Design activity management, taking into account these factors, is aimed at the effective use of positive and compensation of negative personal features of people involved in the design.

The need for such management is due to:

- the nature of the problems associated with the creation of new technical systems (their solution is possible with the appropriate concentration of the abilities and qualifications of designers in the most important areas);
- the scale, diversity and multi-stage character of design processes (they are usually carried out by large teams over a long period of time and require planning, monitoring and coordination);
- the responsibility of designers for hidden characteristics of the state of the created technical system: design and construction flaws, inoperable procedures, etc., which, in combination with obvious failures and other initiating events, create the possibility of an accident (this requires monitoring these processes, checking their results, implementing various preventive and corrective measures).

In practice, several levels of design activity management have been formed; they are interrelated, but each is characterized by certain goals, ways to achieve them, and available resources. The highest level is the state one: it is associated with taking into account the effects of various types of technical systems on public safety. Management at this level is aimed at preventing the creation and commissioning of not entirely reliable technical systems that may pose a threat to society. This goal is achieved primarily through licensing of design activities. Licensing acts as a "threshold filter" for the qualification of designers, carrying out the "culling" of insufficiently competent organizations and the products they create.

The state-level management of design activities also manifests itself in the standardization of a number of their aspects. Thus, the Ministry of Labor and Social Protection of the Russian Federation (Russian abbreviated designation is Mintrud) approves professional standards containing requirements for the qualifications of employees that are necessary to perform labor functions. According to the *Labor*

Code of the Russian Federation,[17] these requirements (currently established for a number of professions related to design) are mandatory for employers to apply. To ensure the correct selection, allocation and use of personnel in organizations, including design ones, it is recommended to use the *Qualification Directory of job positions for managers, specialists and other employees* approved by Mintrud.

The Federal Agency for Technical Regulation and Metrology (Rosstandart) has approved and put into effect standards for design management and quality management in design, which are applicable to any design activity (in accordance with the Federal Law *On Technical Regulation*, they relate to documents of voluntary application).

The influence of the state on the possibilities of creating socially useful and reliable technical systems is also maintained by supporting the competitive environment, higher technical education and scientific research, promoting the sustainability and development of industries important for safety.

When licensing and carrying out other measures of state regulation of design activities, the specifics of industry branches are taken into account. The sectoral (branch-oriented) management principle was characteristic of the USSR industry, when a certain ministry was at the head of each branch. Currently, this principle remains valid in important sectors of Russian industry, but it is implemented in state and united corporations, which include design, construction, engineering and other organizations as well as industrial enterprises. Licensing, which reflects this principle and, consequently, the branch-oriented level of design activity management, is discussed below in Sect. 3.3.2.

In the documents regulating safety in industries of national importance, the influence of human factors on the parameters of technical systems is being recognized at all stages of their life cycle, including the design stage. For example, according to the basic regulatory document of Russian nuclear power industry (NP-001-15), to "every employee or organization involved in the placement, construction, operation and decommissioning of nuclear power plants, design and manufacturing of their … elements, the safety culture shall be instilled and maintained".

A lower level is the management of design activities in design, construction, engineering and other organizations that perform project work (within a certain industry branch or independently of it). The organization-level management reflects the state regulation, the design tasks facing the organization and ways to solve them by teams and individuals in accordance with the culture inherent in the organization. This level is considered in Sect. 3.3.3. When considering, the lowest level in the hierarchy is also affected—the personal level, which is perhaps the key one in relation to the manifestations of human factors.

[17] *Labor Code of the Russian Federation*. The Federal Law No. 197-FZ of 30.12.2001 (as amended on 24.04.2020).

3.3.2 Licensing

Licensing serves, as described in Sect. 3.3.1, to manage design activities on the part of the state and is applied to industry branches important for safety, to confirm that the qualification of designers meets the problems inherent in the creation of technical systems, including problems that arise in conditions where large teams work for a long time. The Russian Federal Law *On Licensing of Certain Types of Activities*[18] (Article 12) provides, among other things, licensing of the development, production, testing and repair of air vehicles and aviation equipment, as well as space activities. The scope and procedure of licensing in the field of aviation technology is determined by the relevant Statute of the Russian Government.[19] Licensing control in this field is carried out by the Ministry of Industry and Trade of the Russian Federation (Minpromtorg). Licensing of space activities, which include, along with other works, the creation, production, testing, repair, modernization and disposal of rockets, spaceships and their equipment, is regulated by another statute.[20] The licensing authority in this area is the State Corporation Roscosmos.

An applicant for a license in one or another of these areas must have (in addition to fulfilling other conditions):

- personnel with professional education who meet the established qualification requirements, including certified specialists;
- production control system necessary for the implementation of the declared activity.

Licensing in Russian nuclear power industry is determined by the Federal Law *On the Use of Atomic Energy*[21] and is carried out by Rostechnadzor. The licenses issued confirm the licensees' right to design nuclear power plants, to develop and manufacture their equipment, to perform works on placement, construction, operation, decommissioning and other activities in the field of nuclear energy use. The license is issued on the basis of an examination of the applicant's documents substantiating the type of declared activity, including an assessment of the adequacy of the regulatory framework used by the applicant, information about his actual activities, the availability and qualifications of personnel and the functioning of the quality management system. The main resources of this method used by the state to control project activities, in particular to implement design activity management on the state level, include the professionalism of the personnel of regulatory authorities and the experts involved by them, as well as the possibility of influencing the qualifications of the licensee by refusing to issue a license, license control, suspension of the license, etc.

---

[18] Federal Law No. 99-FZ of 04.05.2011 (as amended on 02.08.2019).

[19] Statute on licensing the development, production, testing and repair of aircraft vehicles and aviation equipment/Approved by the Decree of the Government of the Russian Federation No. 240 dated 28.03.2012.

[20] Statute on licensing the space activities/Approved by the Decree of the Government of the Russian Federation No. 160 dated 22.02.2012.

[21] Federal Law No. 170-FZ of 20.10.1995 (as amended on July 3, 2016).

Licensing is intended, as already noted above, to ensure the necessary level of qualification of people who create technical systems that affect public safety and to prevent the degradation of technologies. This control method is used in all safety regulation systems in the world. Along with the obvious advantages, it also has some limitations. The latter are determined, for example, by differences in licensing in different industry branches. Licensing requirements established for the construction of airplanes and helicopters, the rocket and space industry or nuclear power are associated with various regulatory legal acts, and their implementation is monitored by various regulatory authorities. There are differences in the qualifications of experts and their interpretation of safety principles. Licensing of activities becomes significantly complicated in the case of the creation of a new type of technical system that combines the technologies of a number of industries, for example, an aerospace vehicle or a spacecraft with nuclear reactor.

Another shortcoming of design activity management through licenses is the largely formal nature of the licensing authority's assessment of information on the qualifications of design personnel and the licensee's availability of a production control system. Behind the facade of the information provided, there may be a loss of actual qualifications due to a long break in the performance of large-scale design work, a change of generations of employees, the retirement of key specialists, etc. Such a situation arises, for example, when the development of industries recognized by society or the government as rather dangerous and unpromising after any resonant disasters (nuclear power, partly cosmonautics) stops for a while. Resumption of activity is possible, but this happens in conditions of lower (compared to continuous development) qualifications of the license applicant. The issues of falling qualifications, reducing the social value of certain industries and the impact of technical policy on them are discussed in Chap. 5.

The formal implementation of licensing is also facilitated by the obvious circumstance that some essential requirements for the knowledge and skills of specialists involved in the development of a technical system can be determined only at the level of a design or construction organization, and only, experienced employees or heads of its thematic divisions are able to formulate them most accurately. Because of this circumstance, formalism is also inherent in the professional standards established by the state.

These shortcomings of the control on the part of the state can be compensated at lower levels of design activity management. For the purpose of such compensation, in particular, periodic certification of specialists in organizations related to industries important for safety (for example, in nuclear energy) is carried out. The content of the certification procedure is the assessment of personnel qualifications in accordance with established indicators and the issuance of appropriate certificates. But, this procedure is also not without formalities.

The connection of licensing with the management of design activities in organizations makes it possible to partially eliminate the noted shortcomings, but cannot completely exclude them. Their existence, in principle, has a negative impact on safety, especially if the designed technical system belongs to a new type.

3.3.3 Management at the Organization Level

The objective function of design activity management at the organizational level, taking into account human factors, includes two main components: the effective use of the professional and creative potential of designers and minimization of their errors. The first component is most in demand when designing fundamentally new technical systems, the development of which is associated with solving non-standard problems; the second is relevant in all cases.

The organization-level management is carried out by the heads of departments in accordance with the hierarchical structure of the organization (see Sect. 3.1.3). When performing the official functions, they give each component a certain weight, depending on the design carried out by the organization, the rules adopted in the industry branch and organization, traditions, current situation, etc. Management resources at this level are the professional and personal qualities of the heads of departments, the measures available to them to organize work, encourage subordinates and improve their skills, as well as influences of administrative or psychological nature. Due to such management, the behavior of individuals is transformed into the behavior of the organization as a whole; it can change over time in accordance with the development of competencies, accumulation of experience, socioeconomic conditions, etc.

The formation of Russian approaches to organizational management, (including at the level of individual organizations), can be divided into the following stages:

- works of A. K. Gastev[22] on scientific organization of labor and management; establishment of the Central Institute of Labor in Moscow (1921–1940);
- projects of the first Soviet five-year plans in the 1930s (application and development of labor management methods);
- projects for the creation of Soviet atomic weapons and intercontinental ballistic rockets, as well as the first space ships (1946–1961);
- development and application of system methods, creation of automated control systems for organizations and factories in various sectors of the Soviet economy (1970s).

The author's work at NPO *Energia* (1975–1994) and NIKIET (in 1990–2002 as a visiting consulting engineer, and since 2005 for about 15 years as a full-time employee) made it possible to get acquainted with the management system in these leading organizations of the rocket and space industry and in nuclear energy.

A notable feature of project management, in particular design activity management, at NPO *Energia* in the design of the *Energia-Buran* space transport system, the *Mir* long-term orbital station and a number of other spacecraft in the last quarter of the

[22] Gastev, Aleksei Kapitonovich, (1882–1939), is Russian revolutionary, trade unionist, poet and writer, theorist of the scientific organization of labor in the USSR and head of the Central Institute of Labor. Unlike Frederick Taylor, who emphasized the system and organizations, and Henry Ford, who focused on improving production techniques, A. K. Gastev focused on human factors.

XX century was the encouragement of initiative developments, including interdisciplinary ones, which contributed to the solution of scientific and technical problems related to these large projects. The hierarchical structure of the "Korolyov's firm",[23] which corresponded to the socialistic way of planning and administrative management, in practice did not prevent business communication between specialists and managers on different levels, cooperation with representatives of other organizations involved in cooperation, promotion of reasonable initiatives and professional development of employees. New methods and technical ideas, after checking them in departments, presenting them in technical reports and approval by scientific and technical councils, immediately went "to the metal". Due to unique nature of the systems being created, the existing regulatory documents partially were losing their mandatory status, and designers were turning to experiments that were not provided for by any standards.

The author sometimes had to hear opinions about the processes of degradation of management after the death of S. P. Korolyov, which also influenced the failure of the N-1 project (see Sect. 3.1.4). But, perhaps, it was this failure which became a bitter experience for the entire team of the "Korolyov's firm", successful in the past, combined with new super-complex tasks and the arrival of V. P. Glushko[24] as general designer in 1974, that led to the strengthening of the innovative aspects in the further management of design activities. Then, these aspects were manifested for more than two decades.

The author's work at NIKIET began in connection with the participation of this institute in the ITER project (see Sect. 3.1.4). The situation in the Soviet nuclear power industry in 1990 was determined by the consequences of the Chernobyl disaster that occurred four years earlier. The impact of this disaster on the personnel of the institute was especially great, since it was this organization which designed the RBMK-1000 reactors installed at the Chernobyl NPP. The first impressions received from acquaintance with NIKIET were strengthened during the subsequent transition there to a permanent job position. Attention in the design and engineering activities of the institute was being focused mainly on safety and minimizing errors through:

- strict compliance of the design procedures with the requirements of regulations and rules;
- large margins of strength (and, consequently, safety) for the designed nuclear equipment;
- enhanced control in the substantiation of strength by analysis, including that with the involvement of third-party specialists who performed comparative calculations;
- in general, a conservative approach to the integrity of the designed structures.

[23] S. P. Korolyev was the Chief Designer of the R-7 rocket described in Sect. 3.1.4 and the first head of the design bureau, then transformed into NPO *Energia*.

[24] Glushko, Valentin Pavlovich, (1908–1989), is one of the founders of the Soviet rocket engine building. He was the general designer of NPO *Energia* from 1974 to 1989 and led the development of the *Energia-Buran* space transport system.

After the Chernobyl disaster, NIKIET did not perform any new large projects, with the exception of participation in ITER (since the beginning of the XXI century, it was gradually narrowing) or development of research reactors which was carrying out for a very long periods. A significant part of the activity was and still is the extension of service life for the existing nuclear power plants with RBMK-1000 reactors.

Comparison of organizational cultures at NPO *Energia* and NIKIET during the periods when the author worked there shows that greater weight in design activity management was given in the first case to the potential for solving problems and in the second case to minimizing errors. Of course, this assessment is quite generalized in nature: it smooths out the heterogeneities inherent in the work of different departments of the same organization.

The noted differences characterize the variation of management in the range including organizations. This variation can be extrapolated to the relevant industry branches. However, design activity management may also depend on the time and nature of the projects.

Thus, after the termination of large Russian space programs at the end of the XX century, there were no breakthrough technical achievements in the activities of JSC S. P. Korolyov Rocket and Space Corporation *Energia* that would require the full realization of the creative potential of its team. The Corporation's participation in international projects (*Sea Launch*, ISS) was limited to a relatively small volume of its own work, and this circumstance has led ultimately to the extinction of the existing organizational culture.[25] In contrast, in 2006, NIKIET started implementing two projects, the novelty and complexity of which led to the use of an unconventional way of design activity management based on the work of project teams outside the hierarchical structure of the organization. These projects were (Spirochkin 2019):

- development of a regulatory framework for equipment and pipeline supports intended for nuclear power plants with WWER type reactors;
- creation of a life cycle information management system for transport reactors.

The first of these projects turned out to be very successful. During the implementation of the second one, significant conceptual advances were made, but the lack of reliable Russian software and powerful hardware did not allow it to be completed properly.

In general, the post-Soviet period can be characterized by the use of approaches to organizational management borrowed from Western management culture.

In Sect. 3.3.1, Russian standards regulating design management were mentioned. They include the documents (GOST R ISO 10006-2005) and (GOST R ISO 21500-2014), identical to the international standards ISO 10006:2003 and ISO 21500:2012.

[25] Probably, not all specialists involved in these projects will agree with this statement. Thus, Sergei N. Atroshenkov drew the author's attention to the fact that a number of elements critical for the existence of the ISS have been created by *Energia*: a motion and navigation control system located in the *Zvezda* service module, jet engines of the Russian segment, as well as main components of the life support system.

Taking into account human factors at organizational management is also regulated by one of the standards of the Project Management series (GOST R 54869-2011).

The methods used to manage design activities in the world are described in many publications, for example, Jones (1982), Pahl et al. (2007), Archibald (2008), PMBOK Guide (2021). In accordance with modern approaches, design work (unlike, for example, current activities of organizations to support the operation stage of technical systems) is carried out by temporary teams and aimed at obtaining unique results. Design activity management in the human dimension includes:

- formation of a design team consisting of employees of the organization and temporarily hired specialists who meet the criteria of necessary competence (education, skills and experience), with the creation of an appropriate organizational structure for the implementation of the design in question;
- creating an environment in which effective design work is possible (building a communication system between its participants, encouraging their skills from the administration, trusting and respectful working relationships in the team, as well as coordinated decision-making and information exchange to meet customer requirements, etc.);
- personnel competence management (training, control and development);
- creating and maintaining a quality culture (customer orientation, staff involvement, systematic approach, continuous improvement, etc.);
- design implementation planning taking into account the limitations inherent in human resources (efficiency, safety, cultural aspects, etc.);
- design implementation control, identification and elimination of inconsistencies;
- managing interaction with the customer, co-executors and other parties interested in the design (clarifying the customer's needs, selecting co-executors, determining the expectations of other parties, identifying and resolving issues important to them, documenting agreements, including informal ones, monitoring the fulfillment of obligations, providing information);
- protection of information, taking into account its confidentiality;
- risk management related to competence, professional responsibility and other personal characteristics of design participants (risk identification, risk reduction, risk sharing, etc.).

When choosing a design team leader, priority should be given to leadership qualities, and when selecting team members, their personal interests, relationships, strengths and weaknesses should be taken into account (GOST R ISO 10006-2005). It is necessary to pay attention to every party interested in the design: individuals, groups or organizations that affect the implementation of the design or to be affected by it. They can be internal or external to the design and have different powers (GOST R ISO 21500-2014). Given the variability of human factors, it is very important for the team leader to take into account the natural limitations inherent in human resources. In particular, the tasks should be distributed according to the capabilities of the design participants (Jones 1982) because "not every task can be executed by every member of the team" (Pahl et al. 2007). This principle cannot be violated without serious consequences for the design development. In addition, when scheduling, it is useful to set floating deadlines for works whose shift does not "jeopardizing the overall lead time" of the development (Pahl et al. 2007).

To minimize erroneous actions of personnel in the design process, as in any other responsible activity, two approaches are used: the person approach and the system one. They differ in causal relationship models and error management methods. The person approach assumes that incorrect actions are caused mainly by deviations in mental processes: forgetfulness, inattention, poor motivation, carelessness, negligence or carelessness (Reason 2000). Counteraction measures are aimed at reducing the undesirable variability of human behavior: posters in the workplace, more detailed regulation of activities, disciplinary sanctions, etc.

The system approach is based on the presumption that humans tend to make errors and errors should be expected even in the best teams. Errors are considered rather as consequences, not as causes; their origin is not so much connected with the "depravity" of human nature, as with systemic factors acting "upstream". These include shortcomings in the workplace and in organizational procedures, a kind of "traps" (Reason 2000).

It was mentioned in Sect. 2.2.4 that high reliability organizations operating potentially dangerous technical systems have distinctive signs: their work is characterized by a high safety culture and resistance to disturbances. A similar culture should be inherent in the organizations that create these systems and should determine the design management. In order to achieve high reliability, design organizations should (along with encouraging the creative potential of employees) develop teams' concern of about possibly adverse events in the operation stage, train personnel to anticipate such events, recognize their sources and ensure the restoration of normal functioning. Instead of performing local fixes, preference should be given to system protection measures.

The errors of the designers of the *Fukushima Daiichi* NPP that caused the accident at this plant have already been repeatedly mentioned above: underestimating the extreme height of the tsunami wave and placing the main and backup power supply devices of nuclear reactors at lower height marks that are jeopardized by possible flooding. Some implicit design errors in airplanes or aerospace ships can serve as examples from another field. One of such errors related to the design of the downstop assembly that restricts the movement of slats in Boeing 737-809 aircraft. A problem in the design of this assembly was discovered during the investigation of a fire that occurred after the landing of a *China Airlines* aircraft at Okinawa Airport (Okinawa-Naha) on August 20, 2007. The immediate cause of the fire was a fuel leak through a hole in the tank wall on the right wing: the leaked fuel came into contact with high-temperature parts of the engine (JTSB 2009). The downstop assembly detached from the aft end and fell out into the track can that housed the inboard main track of the slat on this wing. When the slat was retracted after the landing, the assembly was pressed by the track against the can wall which was adjacent to the fuel tank wall and pierced both walls. Investigators of the accident found a significant error in the design of the downstop assembly: its integrity was ensured only by a washer, which was supposed to prevent the assembly (together with the nut) from falling off from the installation location. In addition, the designers did not take into account that checking the integrity of the assembly during repair or maintenance works was very complicated, since the washer and the nut were located in a place difficult to access. The technician who disassembled and reassembled the downstop at Taipei Airport (Taipei-Taiwan) about a month and a half before the accident could not visually detect that the washer had fallen off and tightened the nut without it. As a result, after some time, the downstop assembly fell out, and this event provoked the catastrophe. Passengers and crewmembers were able to evacuate at the beginning of the fire, and there were no fatalities, but the plane, worth about $ 70 million, burned to the ground.

Even more hidden was the error of the designers of the *SpaceShipTwo* (SS2)—an aerospace vehicle developed by *Scaled Composites* LLC. This error manifested itself during a test flight on October 31, 2014. The crew consisted of a commander and a copilot. The design mission included:

- release of SS2 from the *WhiteKnightTwo* carrier aircraft at an attitude about 15 km;
- acceleration of SS2 from subsonic speed up to 1.4 M[26] in a time interval of 23–26 s with transition to almost vertical flight and reaching apogee (near the conditional boundary of the atmosphere and space);
- reentry with vehicle configuration that provides its aerodynamic braking;
- subsequent glide and landing.

In order to use aerodynamic braking, it was necessary to unlock a feather system (that rotated a feather flap assembly with twin tail booms) after reaching the speed of 1.4 M. The crewmembers were aware that problems could occur during acceleration, due to which the entire mission could be canceled, so they were tense. The unlocking was the responsibility of the copilot. He was apparently afraid not to have time to meet the compressed time interval. Under the pressure of time and under conditions of vibration and loads that he had not experienced recently, he unlocked the feather system when reaching a speed of 0.8 M. Due to the opening of the locks and the rotation of the feather flap assembly at transonic speed, aerodynamic forces arose, that exceeded the design limits, and this led to the catastrophic destruction of the vehicle. As a result, the copilot was killed, and the commander was seriously injured.

The investigation showed (NTSB 2015) that the disaster was caused by the inability of *Scaled Composites* to protect SS2 and, accordingly, the mission from human error. The operational documentation indicated the danger of late unlocking of the feather system, but not early, since such a danger was considered "well-known". Adequate understanding by test pilots of the corresponding risk was not provided. Therefore, the crewmembers had a desire to unlock the feather system as early as possible, so as not to cause the cancellation of further acceleration of the ship due to possible malfunctions. The designers from the *Scaled Composites* did not provide the ship with a system for informing about the achievement of a safe speed to unlock the feather system or protect against premature unlocking, relying entirely on the qualifications of the pilots.

---

[26] M designates the Mach number; in this context, it is the ratio of the velocity of a body in a gaseous medium to the local speed of sound in this medium.

Of course, it is hardly possible to foresee when designing a technical system all adverse initiating events in the stage of operation which can develop into an emergency situation. Nevertheless, it is necessary to target designers to identify "weak points" in the design or operational processes being developed that potentially threaten or contribute to dangerous development scenarios. Similarly, it is necessary to strive for the formation of safety barriers that could protect against possible accidents or catastrophes.

Chapter 4
Changes in Design Due to Development of Views on Safe Human–Machine Interaction

4.1 Need for Changes

4.1.1 Development of Views on Safety

The human desire for greater safety is natural because safety is an integral part of the idea of a comfortable life. In recent decades, there has been a noticeable global trend of tightening safety requirements for the technical systems used (see Sect. 1.2). Obviously, this trend can be explained by the expansion of the technosphere to all regions of the planet, an increase in the number of "second nature" objects, an increase in their power, complexity and interconnectedness. Along with the growth of human population, the global character of economic and social processes is getting stronger, and the price of human life and health is rising in the public opinion not only in economically developed countries but also in developing ones which master the "Western standards" of well-being. Accidents and catastrophes in energy and transport systems or at civil infrastructure facilities are accompanied, thanks to the mass media, by an increasingly strong resonance in people's minds. This, in turn, stimulates public control and influences the technical policies pursued by governments and multinational corporations. A significant part of the causes of emergency events is related to imperfections of technical systems, which are negatively affected by complexity and uncertainty; human factors make a significant contribution to this causal relationship in all stages of the life cycle of the systems, starting with design.

Corrections of design errors and tightening of safety requirements (in the form of changes in relevant regulations, testing of operated systems for increased loads, i.e., conducting "stress tests", strengthening supervisory functions) occur mainly after an accident or catastrophe that reveals hidden flaws in the technical system. In order to minimize risks for newly created systems, the lists of external and internal adverse events are actualized, the parameters of these events are clarified and the principles and methods of design are improved.

However, such changes, which give the technosphere as a whole the necessary stability, no longer correspond to modern views on safe human–machine interaction.

Y. Spirochkin, *Human Factors and Design*,
https://doi.org/10.1007/978-981-19-8832-5_4

They leave open the question: whether all potentially dangerous factors are identified? It is possible that not all situations contributing to their identification took place during the operation of existing technical systems or were the subject of analyzes and experiments. The following question is also reasonable: how adequate are the mathematical models used in the design, including cases of going beyond the experimentally substantiated areas of application? After all, if the modeled object has nonlinear properties, then the problem of correct extrapolation is inevitable.

The development of views on safe human–machine interaction is associated with the transition from Safety I (the level of safety provided by the quality of performance and response to deficiencies identified as a result of accidents and catastrophes) to Safety II (the level of safety associated with the anticipation of emergency events and readiness to manage them). This transition was discussed in Sect. 2.2.5 in relation to the management of operational personnel for safety purposes. It is obvious that such a transition is also necessary in the management of design activities. An important part of the design should be to give the technical system under development the properties of stability in emergency situations and self-healing, which is covered by the concept of Resilience engineering.

Such development of views on safety determinates the need for changes in design as a specific area of human activity, as well as in the results of this activity, so to say, from "external" side. However, there are also "internal" motivating reasons caused by the imperfections of existing approaches to human–machine interaction identified in Chap. 3.

4.1.2 Gaps in Accounting for Human–Machine Interaction

In Chap. 2, the existing approaches to human factors manifested in interaction with technical systems were overviewed. In Chap. 3, these approaches were evaluated from the standpoint of a "generalized designer" who should take such interaction into account when creating systems. In the complex of technical and organizational measures aimed at ensuring the safety of transport systems, energy systems and infrastructure facilities, which is in the whole reliable, gaps were identified—they are associated with insufficient consideration of the negative aspects of human factors and the hazards caused by them.

The identification of such hazards, along with the accumulation of operational experience, the analysis of emergency situations and changes in the assessment of acceptable risks by society, contributes to the development of views on safe human–machine interaction. Now, the opinion is becoming more and more dominant in society that potential hazards should be taken into account regardless of their probability (estimated, as a rule, not quite objectively), and they should be compensated for when designing new technical systems.

In Sect. 3.2.7, it was noted that the variability of personnel actions, including inaction, should be reflected in regulatory documents and design bases. The same should be done with regard to the possibility of an emergency situation when the personnel suppress the functioning of safety barriers (due to negligence or by mistake), for example, a crewmember turns off the automatic control systems of an aircraft. Insufficient to ensure safety is the level of regulatory requirements for the design of technical systems (for design and construction solutions) that must take into account the active behavior of the public in danger conditions. There are no regulatory provisions aimed at excluding the possibility of controlling the system by an attacker (including one from the personnel). Currently, means are being used to block unauthorized control attempts based on checking the criteria for authorized access. Nevertheless, under certain circumstances and with certain efforts, they can be overcome. In addition, without actual recognition of malicious behavior, it is impossible to block the actions of an operator (for example, an aircraft crewmember) who is obsessed with dangerous intentions. Information and mathematical models of human interaction with technical systems in conditions of malicious actions are insufficiently developed.

The existing relationship between active and passive safety means in civil aircraft does not fully meet modern views on safe human–machine interaction. Not all available methods to ensure the survival of occupants of civil airplanes or helicopters in crashes are implemented. A compromise between safety requirements and the possibility of their technical satisfaction is achieved in the design stage under strong pressure from the economy. This commonly made "trade-off" is noted, for example, in the paper Shanahan (2004). Economic considerations are important when creating any technical systems; they determine, in particular, the increase in their dimensions, since this increases the profitability of operation. However, with the increase of dimensions (and, accordingly, the energy accumulated in the system), safety indicators decrease. For nuclear power facilities, this dependence is described by a 3rd-order hyperbole: with a 10-time increase in power, integral safety indicators decrease by 1000 times, as stated in the publication Chumak and Schepetina (2015) with a link to Hattori (1997).

The tightening of safety requirements that is currently taking place has an impact on shifting this compromise toward less risk. In nuclear power industry, for example, the concept of small modular reactors is being developing and implementing, that meets greater safety by physical and mechanical criteria. In general, the need to improve safety requires changes in the design paradigm, including an adjustment of the relationship between safety characteristics and economic indicators in the design objective function. The principles of managing rapidly developing critical situations should also be changed (which is especially important for aircraft). Management improvement can be achieved by strengthening the role of automatic control (capable of recognizing hazards in a timely manner and taking the necessary measures to eliminate danger or rescue aircraft occupants) and reducing the dependence of safety on the actions or inaction of personnel (crew). There are insufficiently resolved issues of human–machine interaction in vehicle crashes. One of them is that the position of the sitting place occupied by a passenger in an airplane may effect on the probability of survival in a crash (see Sect. 2.4.5). Similar differences in safety levels in emergency impacts were also found for passenger cars and buses. These effects should be analyzed and compensated for when designing.

There is also a need to improve the criteria for human tolerance of potentially dangerous effects in accidents, which are used in the design of technical systems, along with taking into account the limit states and analysis of the processes of destruction of their elements (Sect. 3.2.4). In the criteria of strength and integrity of the structure, safety factors (or safety margins) are used to guarantee the established level of safety at variations of loading parameters and bearing capacity within statistically substantiated ranges. In the criteria of human tolerance, such coefficients are not accepted, apparently, they, are not even established. This circumstance causes uncertainty in the level of safety (or survival) in accidents, given the significant stochastic variation in the physical characteristics of people and the conditions of accidents, as well as in the trajectories of destruction processes (Sect. 3.2.5).

Possible changes in the paradigm and principles of designing technical systems aimed at safer human–machine interaction are presented below in Sect. 4.2.

4.1.3 Challenges for Design Activity Management

In order to fill the gaps found in accounting for human–machine interaction, and thus ensure the creation of safer technical systems, improvements in the field of design are needed, including regulation, design principles and applied engineering methods. In the conditions of inertia inherent in regulatory documents with respect to their changes, the functions and parameters of the system under development corresponding to the updated safety concepts can first be established in its design basis (after verification by operational practice, they can be transformed into regulatory provisions). When determining the design basis and the subsequent search for design and construction solutions, problems of uncertainty and complexity inevitably arise.

Uncertainty manifests itself when choosing the functions and values of the system parameters that should ensure, along with the efficiency of operation, a specified level of safety in all operating modes, including difficult-to-predict emergencies associated with rare impacts of extreme intensity. Herewith, as discussed above, in Sect. 4.1.2, some issues of safe human–machine interaction have not been sufficiently studied and resolved.

Mathematical models and simulation techniques based on numerical methods of mathematical physics and operations research, which are used for more accurate and effective substantiation of decisions made by designers, contribute to the complexity of design. Due to the possibility of modeling the structure and simulating the behavior of the created human–machine system with any degree of detail in all operating conditions, such models are currently considered as a substitute for physical experiments. However, when trying to use complex mathematical models and simulation techniques in the early stages of design, when detailing is not based on sufficient certainty of the properties of the system, its functioning processes and interaction of elements, the design technology as a step-by-step process of gradual development and complication can be severely deformed. Moreover, this process shall be accompanied by a human understanding of the problems and testing of new ideas with the help of necessary physical experiments. Otherwise, uncontrolled errors may occur during the design. In the subsequent design stages, the algorithmic nature of numerical mathematical models used in simulations leads to a kind of paradox: despite the large volumes of information produced, it is not possible to obtain any logically substantiated functional dependences of safety indicators on structural, operational, or other factors. Numerical mathematical models produce data, moreover, polluted by "noise", but not the knowledge needed by developers. Logically substantiated dependencies, embodying knowledge, can be obtained only taking into account a certain context, and the latter is not yet covered by machine modeling methods. In general, knowledge originates from physical experience; the possibilities of computational methods in this respect are limited. The problem of the correct identification by a designer of dependencies obtained as a result of simulation using numerical models is also complicated by "information overload" (excessive data volume) and the human ability to find non-existent patterns in random noise (Silver 2012). As a result, the computational substantiations of design and construction solutions that claim to be accurate do not always turn out to be reliable.

The uncertainty and complexity inherent in the design process, including those "aggravated" by the inadequate use of numerical mathematical models, and the corresponding errors in design and construction solutions caused by human factors, require a systematic approach. Along with shortcomings in the workplace and in organizational procedures, they form "traps" (see Sect. 3.3.3) and are, accordingly, challenges for the management of design activities. Herewith, management aimed at minimizing errors should be carried out mainly at the level of the organization directly involved in the design process. In industry branches are important for safety, regulation and control at the levels of the state and the branch are also necessary.

Some possible changes in the management of design activities in order to minimize errors related to uncertainty and complexity are discussed below in Sect. 4.3.

4.2 Possible Changes in the Design Paradigm and Principles

4.2.1 The Design Objective Function

Objective function is a term used in optimal design, and such a function can be formulated explicitly and precisely only for relatively simple cases related, for example, to the improvement of certain structural characteristics or some elements of the designed system for specified operating conditions. The optimal design problem is considered as a search for the values of the design variables (e.g., structural parameters) $\{p\} = \{p_i\} = \{p_1, p_2, \ldots p_I\}$, at which a given objective function $\Psi(\{p\})$ reaches its extremal value, for example, the minimum:

$$\Psi(\{p\}) = \min, \tag{4.1}$$

subject to constrains:

$$G(\{p\}) \leq 0; \;\; H(\{p\}) = 0. \tag{4.2}$$

The value of each design variable is searched in the range:

$$p_{i\ \min} \leq p_i \leq p_{i\ \max} (i = 1, 2, \ldots I) \tag{4.3}$$

where $p_{i\ \min}$ and $p_{i\ \max}$ are the allowable limits of variation associated with technological and biological constraints (when considering human–machine interaction) or with uncertainty ranges inherent in the design problem being solved;

I is the number of design variables involved in optimization.

The objective function can represent:

- the mass of a system component (mass optimization inherent in the design of aircraft);
- a certain characteristic of the shape of a structure (corresponding, for example, to greater stability of a pressure vessel and associated with minimizing material consumption);
- the number of components (e.g. the number of pipeline supports, which refers to minimizing production and operating costs),[1] and so on.

When optimizing the crashworthiness of a vehicle, typical objective functions are the maximum acceleration acting during impact, the energy absorption capacity of structural parts, or the human injury indicators (Ambrosio et al. 2001).

Constraints (4.2) are formed based on the criteria accepted in the design. These criteria may include the sets presented in Fig. 3.5 which relate to the strength and integrity of the structure (A), the human tolerability of potentially dangerous effects (B) and their intersection (A & B). In a number of optimization methods, constraints are included in the objective function, so some of these criteria can be taken into account using either formulas (4.2) or expressions (4.1).

The concept of "objective function" is also used in an extended, engineering sense, regardless of the degree of optimization, including a situation where the problem cannot be formulated strictly mathematically—to evaluate the quality of design, compare design variants by several indicators and choose the best one, or determine the directions for finding feasible solutions. The main indicators in the evaluation and comparison relate to the performance of the designed system (achievement of project goals), safety and economic costs. Performance indicators can be the generated electric power for a nuclear power plant, the payload mass for a space transport system, the number of passengers for a civil airplane, etc. To evaluate safety, the following indicators are applicable (Spirochkin 2019):

---

[1] The author participated in the 1990s in the works of the German company KED (Kerntechnik, Entwicklung, Dynamik) on optimizing the shape of containers intended for storing liquid industrial waste, as well as on developing software for optimal placement of pipeline supports.

- the estimated probability of an emergency situation (accident or catastrophe) during operation;
- values of mechanical reliability of elements important for safety, and associated safety factors, safety margins or reliability factors[2];
- number of protection levels;
- the response time of protective devices, their resistance to extreme impacts, the ability to absorb energy, etc.

As indicators of economic costs, designers can consider their total value, components of this value distributed by stages of the life cycle of a technical system, production cost, payback period, etc.

The corresponding sets of indicators can be designated as $\{\mathcal{P}\}$ (performance indicators), $\{\mathcal{S}\}$ (safety indicators) and $\{\mathcal{E}\}$ (economic cost indicators); they all depend on the sought design parameters $\{p\}$. The required safety indicators related to normal operation conditions differ from those that correspond to survival in an accident; accordingly, the costs of ensuring safety in these situational categories are also different. As a rule, the loading conditions in case of an accident are not at all those that are typical for normal operation. Moreover, even in different modes of normal operation, the same structural element can be subjected to loads that differ significantly in maximum values or time profiles. For this reason, the optimal parameters of structural elements obtained according to the criteria for certain conditions turn out to be not optimal in relation to another situation. It is also logical to conclude that technical systems highly optimized for normal operation can be very vulnerable in accidents, and designers should take appropriate compensatory measures to ensure safety.

The situation becomes even more complicated if different sets of indicators have the opposite effect on the generalized objective function Ψ used when assessing the quality of the design under consideration (in this case, its maximum is sought). It is reasonable to digress from certain values of $\{\mathcal{P}\}$, $\{\mathcal{S}\}$, and $\{\mathcal{E}\}$ (which are subject to agreement between the Customer and the Contractor of the design and are set in the corresponding technical specification) and consider the fundamental dependence $\Psi = \Psi(\{\mathcal{P}\}, \{\mathcal{S}\}, \{\mathcal{E}\})$. The increase in Ψ is facilitated by an increase in performance indicators, e.g. the power of the technical system (energy accumulated in it) and a reduction in economic costs, but this is fraught with a deterioration in safety indicators (see Sect. 4.1.2). The values of Ψ increase with increased safety, but for greater safety, it is usually necessary to raise costs and reduce the power of the system, which lowers the objective function. Thus, the influence of indicators $\{\mathcal{P}\}$ and $\{\mathcal{E}\}$ on the objective function is opposite to the influence of $\{\mathcal{S}\}$, and therefore any design and construction solution that is considered as the best according to the first two sets of indicators turns out to be not the best, and sometimes just bad in relation to the third set. As a result, only a compromise is possible between operational performance and costs, on the one hand, and safety characteristics, on the other; in practice, it is implemented at a certain relationship of $\{\mathcal{P}\}$, $\{\mathcal{E}\}$ and $\{\mathcal{S}\}$, which is considered acceptable.

[2] These terms denote the same parameter or similar parameters used in different branches of technology.

In order to simultaneously take into account several indicators that also have a multidirectional effect on the objective function, a combination of them is usually built. For example, considering only one indicator in each of the sets, it is admissible to use a weighted sum:

$$\Psi = w_{\mathcal{P}}\overline{\mathcal{P}} + w_{\mathcal{S}}\overline{\mathcal{S}} + w_{\mathcal{E}}\overline{\mathcal{E}} \tag{4.4}$$

where the weighting coefficients $w_{\mathcal{P}}, w_{\mathcal{S}}, w_{\mathcal{E}}$ reflect the significance of $\{\mathcal{P}\}$, $\{\mathcal{S}\}$, and $\{\mathcal{E}\}$ in the design, respectively; meanwhile

$$w_{\mathcal{P}} + w_{\mathcal{S}} + w_{\mathcal{E}} = 1 \tag{4.5}$$

and the dimensionless indicators are determined by the formulas:

$$\overline{\mathcal{P}} = \frac{\mathcal{P}}{\mathcal{P}_{\max}}; \overline{\mathcal{S}} = 1 - P_A; \overline{\mathcal{E}} = 1 - \frac{\mathcal{E}}{\mathcal{E}_{\max}}. \tag{4.6}$$

Here $\mathcal{P}_{\max}$ is the maximum value of the performance indicator considered as the most important; $\mathcal{E}_{\max}$ is the maximum value of economic costs; P_A is the probability of an emergency situation accepted in the design. Taking into account that the values of P_A are usually very small compared to 1 and inconvenient to use, $\overline{\mathcal{S}}$ can alternatively be expressed as

$$\overline{\mathcal{S}} = \frac{lg(P_A)}{lg\left(\hat{P}_A\right)} \tag{4.7}$$

where $\hat{P}_A$ is the corresponding probability value established in regulatory documents. As a rule, it is equal to 10^{-6} for an accident and 10^{-7} for a catastrophe.

If all indicators are equally significant, then $w_{\mathcal{P}} = w_{\mathcal{S}} = w_{\mathcal{E}} = {}^1/_3$. If, for example, a greater weight in the design is assigned to performance or cost indicators, then according to (4.5) this entails a decrease in $w_{\mathcal{S}}$. It means that the contribution of safety characteristics to the objective function is reduced and they have less influence on the evaluation of the quality of design and the choice of the best design variant.

Optimization is carried out, as a rule, numerically; during this process (usually in a step-by-step mode), the coefficients of sensitivity of the objective function to small changes in the design variables are calculated. The optimization algorithm uses these coefficients to determine the direction in which the optimal point should be searched:

$$\frac{d\Psi}{dp_i} \cong \frac{\Delta\Psi}{\Delta p_i} = \frac{\Psi(\{p\})|_{p_i+\Delta p_i} - \Psi(\{p\})|_{p_i}}{\Delta p_i} (i = 1,\ 2, \ldots I). \tag{4.8}$$

If the objective function is represented by the weighted sum (4.4), then the sensitivity coefficients can be calculated by the formula:

$$\frac{d\Psi}{dp_i} = w_{\mathcal{P}} \frac{d\overline{\mathcal{P}}}{dp_i} + w_{\mathcal{S}} \frac{d\overline{\mathcal{S}}}{dp_i} + w_{\mathcal{E}} \frac{d\overline{\mathcal{E}}}{dp_i}. \quad (4.9)$$

When the weighting coefficient $w_{\mathcal{S}}$ decreases, the sensitivity of Ψ to safety indicators (corresponding to the second term in this formula) falls. In this case, the optimization algorithm may prefer at a certain step a search direction that will not meet the required level of $\overline{\mathcal{S}}$. In addition, the small value of $w_{\mathcal{S}} \frac{d\overline{\mathcal{S}}}{dp_i}$ may impair the accuracy of the numerical determination of the optimal point in regard of $\overline{\mathcal{S}}$, and this may become critical near the boundaries of safety.

Instead of the linear combination (4.4), a weighted quadratic sum of indicators can be used:

$$\Psi = \sqrt{w_{\mathcal{P}}^2 \overline{\mathcal{P}}^2 + w_{\mathcal{S}}^2 \overline{\mathcal{S}}^2 + w_{\mathcal{E}}^2 \overline{\mathcal{E}}^2}. \quad (4.10)$$

In the special case when $w_{\mathcal{P}} = w_{\mathcal{S}} = w_{\mathcal{E}} = 1$, the quadratic sum represents the module of the vector of indicators.

If it is necessary to take into account several indicators in each of the considered sets $\{\mathcal{P}\}$, $\{\mathcal{S}\}$, and $\{\mathcal{E}\}$, the objective function can be constructed in the form of various other combinations in accordance with the specifics of the task.

Safety of a technical system is determined by the reliability values of its main components:

- structure, R_s;
- automatic systems (control system, active and passive safety systems), R_a;
- humans (personnel), R_h.

When calculating the safety indicator related to the probability of emergency situation P_A caused by the failure of the listed components, and provided that they are independent of each other, the expression is applicable:

$$\overline{\mathcal{S}} = 1 - P_A = 1 - (1 - R_s)(1 - R_a)(1 - R_h). \quad (4.11)$$

Modern methods of design, manufacture and maintenance of technical systems provide R_s and R_a values at very high levels: 0.9(n) (where n is the number of nines; usually it is greater than 5); they exceed the statistically substantiated values of R_h (RB-100-15). The decrease in the safety indicator due to the insufficient reliability of humans, R_h (caused, for example, by their errors) can be compensated for by the greater reliability of automatic systems, R_a, but their functioning must be protected from malicious interference. The possibilities of such compensation are limited by the very presence of the multiplier $(1 - R_h)$, which characterizes the negative influence of human factors. On the other hand, the reliable functioning of humans in the role of personnel is able to minimize deviations in the behavior of the technical system caused by a failure of automatic controls. The ambiguity of human influence demonstrated by expression (4.11) does not allow, at least at this stage of technology development, to completely exclude the human operator from the control loop. The search for the optimal combination of automatically performed functions and human participation in the control of technical systems remains an urgent problem.

The above considerations illustrate the complexity of optimizing technical systems being created and some ways of constructing a generalized objective function of design taking into account a variety of criteria and indicators, including those that contradict each other. At the same time, they show the possibility of shifting the design paradigm in order to achieve greater safety of human–machine interaction. Such a shift, which seems necessary to the author, corresponds to a change in the relationship between economic costs and safety characteristics of the technical system in the objective function of design in the direction of increasing the weight of safety. Simple mathematical expressions representing the contradictory influence of the human factors on safety (as a component of the objective function) also serve as substantiation for further shift. The next-generation technical systems should be able to recognize critical situations that may arise during operation due to the insufficient reliability of personnel (taking into account the variability of its actions and possible malicious intentions), and minimize hazards through appropriate functioning of advanced automatic controllers (including measures to ensure rescue).

4.2.2 In-Service Recognition and Control of Critical Situations

It is shown above that the insufficient reliability of the personnel (caused by the variability of its actions including errors and negligence), as well as possible malicious intent (for example, when terrorists gain access to control the system), cannot be fully compensated by the high reliability of other system components (mechanical, electrical or electronic) created within the existing design paradigm. This paradigm provides for the use of an a priori model of operation based on statistical data for similar technical systems. The question of whether the available statistics cover all possible operation situations and all the features of the system's behavior, including rare impacts of extreme intensity, especially unpredictable ones that may occur during terrorist attacks, and the response to them, remains open (a similar question was put in a more concise form in Sect. 4.1.1). The mechanical reliability of the system structure, when the operating conditions change within uncontrolled ranges and the loads acting on the structural elements exceed the strength limits considered in the design, is also uncertain.

The traditional approach to design does not actually ensure safety of technical systems in relation to unpredictable events in general and terrorist attacks in particular. In recent years, some efforts have been made to change this approach. In the publication Makhutov (2009), it was proposed to take into account the possibility of terrorist attacks or other negative manifestations of human factors using probabilistic models, similar to models describing the risks of natural disasters or technology-induced catastrophes. According to this proposal, the development of a critical situation caused by inadequate human–machine interaction should be considered as a probabilistic process in which dangerous conditions, in particular limit states in structural elements, are reached at certain time moments, but unlike the "regular" emergency situation provided for in existing regulatory documents, this occurs at a higher rate. When assessing the probability of such a development, not only the actions of an attacker who penetrated from the outside should be taken into account, but also the actions of the technical system personnel, as well as the behavior of security staff or decision makers at various levels.

In the publication Dédale (2013), on the contrary, the inapplicability of statistics for predicting of low probable extreme impacts on technical systems is substantiated. Interpreting impacts of any intensity as deviations from some average values and approximating them with Gaussian (normal) distribution does not take into account complexity of the real world and actually equalizes risks of different origins. For these reasons, it is proposed to shift the traditional design paradigm based on such a priori model of future operation in the direction of creating systems resilient in relation to unpredictable events. However, no applied design methods implementing such a concept are presented.

While a generally accepted alternative to probabilistic-statistical accounting of the variability of human actions, including their extreme (terrorist) manifestations, has not yet been found, more adequate distributions than Gaussian, for example, the power law, are used, and the Bayesian approach is also applied to refine a priori probabilistic estimates (Silver 2012). The probability distribution density of a variable x in accordance with the power law is expressed by the formula:

$$f(x) = cx^{-\alpha}. \tag{4.12}$$

Such distribution with the exponent $\alpha = 2.5$ is shown graphically in Fig. 4.1.

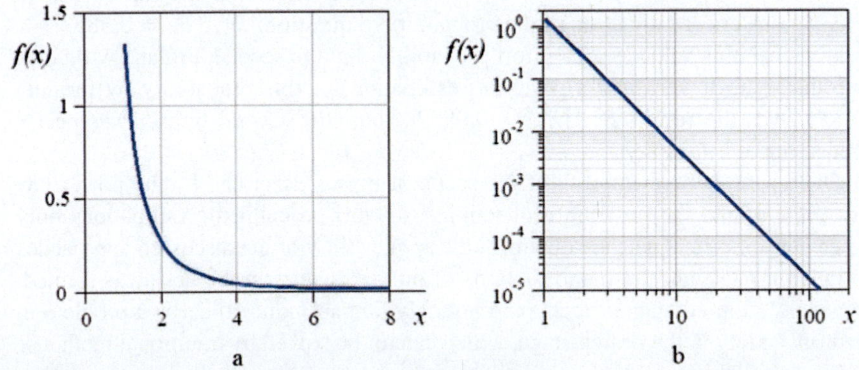

Fig. 4.1 Example of the power law distribution with exponent $\alpha = 2.5$: **a** in ordinary coordinates, **b** on logarithmic scales. Based on data from Newman (2005)

A variable obeying the power law distribution has well-defined mean value only if $\alpha > 2$, and meaningful variance or standard deviation if $\alpha > 3$ (Newman 2005). Many of the studied natural phenomena are characterized by the values of exponent in the range: $2 < \alpha < 3$, therefore there is a possibility, they may occur in the form of a Black Swan (Taleb 2007), i.e., unexpected events developing in an unpredictable scope. In particular, terrorist attacks with more damage and casualties fatalities than occurred before are likely in the future, but it is impossible to predict exactly where and when a new attack will occur.

The Bayesian approach, which allows "processing" inaccurate, even subjective, probabilistic estimates, was briefly presented in Sect. 3.1.4. An a priori estimation of the probability of a rare event A, $P(A)$, in the absence of reliable statistics expresses the degree of confidence that such an event will occur. According to the Bayes' theorem, this estimation could be refined in the case of obtaining information related to the occurrence of a certain event B, i.e., in the form of a conditional probability:

$$P(A|B) = \frac{P(A) \times P(B|A)}{P(A) \times P(B|A) + P(B|not A) \times [1 - P(A)]} \tag{4.13}$$

where $P(B|A)$ is the probability of event B associated with the fact that event A occurred (conditional probability);

$P(B|not A)$ is the probability of event B provided that event A does not occur.

To illustrate the application of this approach, we reproduce an example given in the book Silver (2012) concerning the terrorist attacks on the World Trade Center on September 11, 2001. These attacks were carried out by two airplanes crashed into the towers of the Center one after the other in a short period of time. A priori estimation of the probability that an airplane deliberately collides with the tower, in this case with the north one, can be assumed very small: $P(A) = 0.5 \times 10^{-4}$. Event B is the first impact itself, i.e. the event that actually occurred, so $P(B|A) = 1$. The probability that an airplane may hit a skyscraper for a random reason can be calculated based on statistics for the previous 25,000 days. There were two such accidents, so $P(B|not A) = 0.8 \times 10^{-4}$. Substituting these data into Eq. (4.13), we obtain a posteriori estimation of the probability in question: $P(A|B) = 0.385$.

If we use this value as an a priori probability that the second airplane will crash into the south tower, then a posteriori estimation for this probability by formula (4.13) gives the result: $P(A|B) = 0.9999$. Thus, the second impact was nearly predetermined.

Due to the unpredictability of the scope of events described by the power law distribution, and the possibility of refining a priori probabilistic estimations only when new empirical data are obtained, the applicability of the discussed approaches to take into account the manifestations of human factors in the design is limited. In general, unpredictable events (and not only human-induced) cause a problem in ensuring safety in the design stage, which cannot be solved by traditional methods.

The unreliability of design specifications based on statistics makes it necessary to estimate the physical parameters of events that go beyond the design basis (in fact, the accepted forecast) directly during the operation of the technical system and respond to them depending on the degree of criticality of the situation, taking into account the available opportunities to minimize the danger. Recall that we are considering here such low probable or unpredictable initiating events that correspond to the negative manifestations of human factors embodied in the human operator of the technical system: his erroneous or malicious actions or his inaction. Since the critical situation can develop very quickly, its management (including assessment, compensation of deviations and measures to save people) is obviously possible only with the help of automatic controllers, the reaction of which exceeds the human abilities. The design tasks should include providing the technical system with such controllers capable of recognizing emerging critical situations that threaten to escalate into an accident or catastrophe and managing them in accordance with the actual operational model, which is determined dynamically at the appropriate time moments or, rather, very short periods of time. Modern views on safe human–machine interaction require endowing the created systems with the properties of readiness for unpredictable events and resilient response to restore normal operation conditions or, if this is not possible, to ensure survival by other means. The implementation of such properties using next-generation automatic controllers is the second proposed shift in the design paradigm; this corresponds to the application of the concepts of Safety II and Resilience engineering (mentioned in Sects. 2.2.5 and 4.1.1) in the field of design and construction solutions.

Such a shift is necessary when designing any potentially dangerous technical systems, but it is most relevant for vehicles, especially aircraft. If the airplanes that crashed into the towers of the World Trade Center had been equipped with electronics capable of recognizing the actions of terrorists and block them, and then turn on the automatic control mode and land both jets at the nearest airfield, the catastrophe (accompanied by the death of almost three thousand people) could have been avoided. Similar automatic control systems would, obviously, prevent airplane crashes caused by suicidal actions of crewmembers (see Sect. 2.1).

Modern airplanes are equipped with automatic systems designed to recognize dangerous situations caused by the proximity of a flying vehicle to another one or to the ground. According to ICAO standards, each aircraft with the mass more than 5700 kg or certified for the transportation of more than 19 passengers, must have an on-board warning system for a possible collision in the air: Traffic Collision Avoidance System (TCAS). This requirement is reproduced in the Russian regulatory document: Federal Aviation Rules, Part *Preparation and Execution of Flights in Civil Aviation of the Russian Federation,*[3] Clause 5.77. Currently, second-generation systems are used: TCAS II, which give pilots not only information about the situation in the air within a radius of 65 km, but also recommendations for eliminating potentially dangerous situations (Traffic Advisory), as well as commands for immediate actions to prevent collisions (Resolution Advisory). These actions are defined in the vertical plane. In the latest version of TCAS II, designated as 7.1 (FAA 2011), (ACAS Guide 2017), the system flaws are fixed which actually contributed to the

[3] Approved by the Directive No. 128 of the Ministry of Transport of the Russian Federation dated July 31, 2009.

collision of the *Bashkir Airlines* Tu-154 M and Boeing 757 (DHL) airplanes over Lake Constance (Bodensee) on July 1, 2002 (ASN 2002). Among the flaws was the insufficiently reliable logic of the system, which did not exclude the possibility of the pilot to control the aircraft contrary to the TCAS commands (e.g., following the instructions of air traffic controller who does not receive and does not monitor such commands).

In accordance with Clause 5.76 of the Federal Aviation Rules (see above), aircraft (of the specified mass or designed for the specified number of passengers) must be equipped with the Ground Proximity Warning System (GPWS). GPWS has the function of assessing the terrain relief in the direction of flight and automatically generating the appropriate audio and visual signals for the flight crew. Now, more advanced systems of the type Terrain Awareness and Warning System (TAWS) are used, which include a database on terrain and obstacles. After the first warning, which draws the pilot's attention to the potential danger of the situation, about 30 s before the calculated collision with the earth's surface, the voice informant gives a signal to increase altitude (FAA-H-8083-6 2021). This signal takes precedence over any other commands including commands from TCAS. Pilots must turn off the autopilot and autothrottle control, remove spoilers if they are released, manually switch the engines to maximum operating mode and begin to gain altitude. The timely and correct reaction of the flight crew lets to avoid a collision with the earth's surface. However, the TAWS function can be manually disabled by the pilot. So did, for example, the commander of the Sukhoi RRJ–95B airplane during a demonstration flight on the island of Java (Indonesia) on May 9, 2012. This flight took place in mountainous terrain unfamiliar to the flight crew. At a certain point, TAWS began to give aural warning: "Terrain–Pull up" and "Avoid Terrain" and corresponding visual signals. The commander inhibited the function of the system, assuming that it was giving incorrect warnings due to a problem with the terrain database and relying on his piloting skills. This assumption proved fatal: the airplane crashed into Mount Salak (KNKT 2012).

In general, at present, despite equipping the aircraft with automatic systems for recognizing critical situations and informing pilots about them, including repeated signals accompanied by clear recommendations, the control actions themselves remain for the pilot. The above-mentioned catastrophes show that it is human presence in the control loop of the machine that limits the possibilities of ensuring safety, at least in situations created by his malicious actions, erroneous actions or inaction and characterized by rapid development.

The exclusion of an "unpredictable" and "unreliable" human operator from the management of a critical situation should be accompanied by the replacement of the functions required of him with "smart" automatic controllers, which should assess the situation and respond properly. Taking into account the significant differences in the branches of technology, it is possible to formulate, without going into details, only a general approach to the design of such controllers.

When recognizing a critical situation and looking for ways to maintain control of the technical system (its functioning process and, accordingly, over the situation) in order to minimize potential losses, it is necessary to analyze:

- intensity of dangerous effects (causing deviations from normal operation conditions);
- the actual state of the system from the point of view of safety (integrity, controllability, deterioration of physical and mechanical properties);
- efforts necessary to minimize the hazards (to compensate for deviations);
- the variability of the situation and the time available for compensation;
- the occurrence of additional effects when compensating, for example, increased inertial loads at a certain combination of time profiles of compensating forces and dynamic properties of the system structure;
- restrictions imposed by the environment (for example, terrain or meteorological conditions);
- additional restrictions caused by the actual state of the system.

The listed aspects, expressed quantitatively, form a set of variables describing the situation, in which the technical system is "located", i.e., the variables of the state of the process of its functioning. This state can be within one of the following possible categories: normal operation, violation of normal operation, accident, or catastrophe.

Estimation of the intensity of physical and mechanical effects or deviations, the "internal state" of the system (from the point of view of safety and controllability), the rate of change of the situation and some imposed restrictions are possible on the basis of in-service monitoring. The values of the variables obtained in the monitoring process should provide, all together or separately, an unambiguous assessment of the state of the functioning process. When the situation becomes critical, i.e., is characterized by the possibility of losing control over it, the values of these variables should provide for the development an algorithm for compensating for deviations and (or) saving people within the available time reserve (also taking into account the time interval required for the implementation of the algorithm).

In-service monitoring can cover two groups of data: the first group includes actual system parameters, applied loads and environmental parameters that are measured directly (by sensors), and the second one comprises data determined indirectly based on measurements of other parameters. For aircraft, the first group may include deflections of aerodynamic control surfaces, engine power (turbine rotation speed), horizontal and vertical components of the flight velocity vector, pitch and roll angles, etc. (they may coincide with the data recorded by the flight recorder). At nuclear power facilities, parameters are monitored, which could characterize:

- technological processes (temperature, pressure, coolant flow rate, etc.);
- concomitant processes (vibration, noise, possible coolant leakage);
- processes related to the application of additional loads (including testing processes, extreme internal and external effects);
- integrity and aging of the structure (actual properties of structural materials; geometry of crucial elements and their connections with other elements, including gaps,

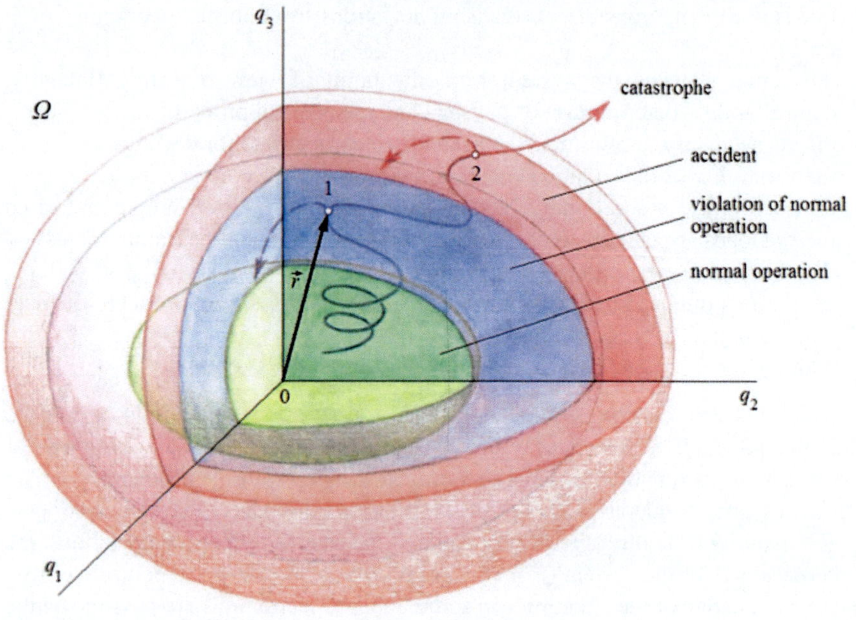

Fig. 4.2 Space of variables, its characteristic areas and the trajectory of state of the functioning process of a technical system

tightness, misalignment; displacements in movable joints and assembly units, localization and sizes of defects and damages).[4]

To recognize critical situations and analyze the possibilities of managing them, the methodology of reliability theory (Bolotin 1971) can be used, as well as its applications to safety issues (IMASh 1992; Makhutov 2009). According to this methodology, operation of a technical system can be represented as a "trajectory" of state of the functioning process occurring in a space of variables $\Omega(q_1, q_2, \ldots q_L)$ over time ($q_1, q_2, \ldots q_L$ are the considered variables of state of the functioning process, and L is their number). Each point of the trajectory is defined by a radius vector $\vec{r}(q_1, q_2, \ldots q_L)$. Figure 4.2 shows such a space and the relevant trajectory for the case of three variables: q_1, q_2, q_3. The space of state variables Ω can be divided into characteristic areas corresponding to the above state categories (normal operation, violation of normal operation, accident and catastrophe).

Suppose that our technical system has a built-in "smart" electronic controller capable of recognizing which area the system is in, based on monitoring the variables of state of the functioning process (both groups of data discussed above) and evaluating them in accordance with established criteria. If the system functions as

[4] Most of this data is recorded, as a rule, when the nuclear reactor is shut down during scheduled periods of operation.

designed and all these variables are within the specified limits, the current situation belongs to the category of normal operation and the corresponding area in Fig. 4.2.

Suppose further that due to erroneous or malicious actions of a human operator or his inaction, the variables of state of the functioning process change in such a way that this process goes from the area of normal operation and falls into the area of violations of normal operation, e.g., into point 1 (see Fig. 4.2). If the recognition of this situation, performed by a "smart" automatic controller, demonstrates that there are still opportunities to restore normal operation (a critical situation of the first degree, potentially reversible), then the actions of the human operator are blocked, and the necessary compensating forces (determined in real-time) begin to act on the system in accordance with the developed algorithm. These forces return the functioning process into the area of normal operation.

If the functioning process falls into the area of accident (which is characterized, for example, by a severe deterioration of integrity or controllability of the system), and this situation is recognized only at point 2 (a critical situation of the second degree, irreversible), then the actions of the human operator are also blocked automatically. But now, our "smart" automatic controller looks for another way to maintain control over the situation and develops a survival algorithm. This algorithm shall be implemented by appropriate safety equipment that prevents the accident from escalating into a catastrophe. For military aircraft, the chosen method of survival may include a controlled emergency landing or ejection of pilot (or pilots). Obviously, the complexity of the problem of recognizing critical situations and managing them dictates the need to create such "smart" electronic controllers that would have artificial intelligence.

When implementing the above approach to the exclusion of human operator from control in critical situations, functional components described in the patents for civil aircraft safety systems (Berestov et al. 2004; Daveze et al. 2009; Berestov et al. 2010) can be applied. Currently, these components only partially cover the range of "smart" (or intelligent) control required, and they are available at the level of schematics and specifications.

4.2.3 Ensuring Survival in Accidents

In an irreversible critical situation, when the integrity or controllability of the technical system is violated, and it is impossible to return the functioning process to the area of normal operation (see Sect. 4.2.2), the survival of people (personnel, passengers, population) who are in danger if the accident escalates into a catastrophe must be ensured. If such a situation is caused by negative manifestations of human factors, e.g., erroneous or malicious actions of human operator or his inaction, the method of survival should be chosen and implemented exclusively by a "smart" automatic controller working in real time with a speed ahead of the human reaction. In complicated and rapidly evolving critical situations of another nature, an intelligent automatic control may also be preferable. The development of the survival algorithm is influenced by the results of the analysis of the situation taking into account aspects that partially coincide with those listed in Sect. 4.2.2:

- efforts required to mitigate the potential consequences of the accident;
- time available for these efforts;
- the occurrence of additional effects when these efforts are made, e.g., additional inertial loads;
- the remaining (as yet available) possibilities to control the system after deterioration of its integrity or controllability;
- restrictions imposed by the environment and the actual state of the system.

The survival algorithm can be developed (or selected from the built-in set) based on the active or passive safety devices that the technical system is equipped with. One of the controversial design principles is the preference (almost exclusive) given to passive safety devices. In air and ground transport, these include energy-absorbing structural elements, shock-absorbing seats and seat belts, as well as other design and construction solutions that can reduce the dangerous effects in accidents to acceptable limits. In nuclear power industry, the priority of passive safety devices based on natural triggering and functioning without additional energy sources is established at the regulatory level (NP-001-15). It should be born in mind that the applicability of such devices is limited to the area of changing the parameters of the system state, in which physical processes that ensure natural triggering and functioning can provide an acceptable level of safety. Erroneous or malicious actions of a human operator or his inaction can lead to such a critical situation in which the use of passive safety devices to prevent a catastrophe may not be enough, and more significant efforts will be required to cope with it without fatalities. Taking into account the limited applicability of the principles of passive safety, military airplanes and helicopters are equipped with ejection seats or escape capsules, and in space technology, the use of active emergency rescue systems is mandatory.

Modern views on safe human–machine interaction, including the requirement to ensure survival in accidents, make the introduction of active safety methods in civil aviation actual. These methods could encompass, for example, rescue capsules for passengers and crewmembers, or solid-fuel engines similar to those used in the launch escape systems (LES or SAS, see Sect. 2.2.3), or soft landing systems for spacecraft, etc. The use of active safety means would also lead to the elimination of unequal chances of survival in crashes of airplanes, cars, or buses, depending on the sitting place occupied by the passenger (for airplanes, such data are given in Sect. 2.4.5). Unequal chances are determined, first of all, by the most likely place of application of impact loads for a fast-moving vehicle, i.e. in its front part. Accordingly, the structural parts of the vehicle that are closer to this place are subjected to more intense loads and are destroyed to a greater extent, and occupants who are there are affected by higher accelerations and other hazardous factors. The corresponding differences in safety levels in different parts of the passenger compartment are very difficult to compensate for through the functioning of passive safety devices. At the same time, active safety equipment, for example, in the form of additional inflatable shock-absorbing pillows or special triggered structural elements that increase local resistance to loads and destruction could solve this problem.

However, the capabilities of passive safety means have not yet been exhausted, especially when they are implemented at a new technological level. Among the modern promising developments in this area that could improve safety of occupants in vehicle crashes, we can mention:

- structural materials capable to absorb kinetic energy due to nanosecond-fast phase change at the physicochemical level (Synbiosys 2021);
- the so-called "metamaterial", which combines an elastic, rubber-like substance with tiny magnets embedded in it and is capable to phase shifting (to significantly increase the amount of absorbed energy) rather, at the physico-mechanical level (Liang and Crosby 2022);
- an "anti-explosion system", which prevents the inflammation of aviation kerosene and its burning, and an "anti-trauma system", which fills the aircraft in a crash with a very dense foam within a very short time (about 2 s), blocking the movement of any objects inside (IdeaConnection 2015).

Of course, these developments need careful experimental verification.

An important principle of designing technical systems that guarantees an established level of safety, taking into account the stochastic distribution of loading parameters and mechanical resistance of structural elements, is the use of safety factors, safety margins, or reliability factors.[5] These factors (margins) are mandatory components in the criteria of strength and integrity of the structure; however, they do not considered when assessing the human tolerance of potentially dangerous effects. But, in the end, it is this assessment that allows us to form a judgment about survival in an accident. As a result, as noted in Sect. 4.1.2, there is uncertainty about the guaranteed level of safety. It is obvious that in order to ensure a certain level of safety, when both the parameters of effects, especially emergency ones, on humans and the characteristics of human tolerance of these effects have significant variations, the safety criteria for human–machine interaction, including those that determine survival in accidents, should be formulated as follows:

$$C_k^{(B)} = \left(\mathcal{T}_k^{(B)} \geq f_k^{(B)} \mathcal{F}_k^{(B)}\right)\left(k = 1, 2, \ldots K^{(B)}\right) \tag{4.14}$$

where $\mathcal{T}_k^{(B)}$ is a measure of human tolerance of the k-th dangerous effect to be included in the set of criteria B (see Fig. 3.5); exceeding this measure means fatality;

$\mathcal{F}_k^{(B)}$ is a measure of this effect itself (it poses a danger to humans directly and, therefore, belong to the set B);

$f_k^{(B)}$ is the safety factor (we will use this term) corresponding to both of the above measures and, and accordingly, to the set B; it is greater than 1.

The dangerous effects on humans in vehicle crashes are impact accelerations, deformations of life volume, the time of the accident development during which it

[5] See footnote 2.

is possible to leave the vehicle, etc. In case of a severe accident at a nuclear power plant, which is accompanied by the destruction of the reactor and the activation of localizing safety systems (in particular, the core melt catcher), safety criteria should take into account radiation levels, the time of taking measures to protect and evacuate the population, etc.

The values of $f_k^{(B)}$ can be determined, like the safety factors for structural elements, based on the data of relevant theoretical and experimental studies. The relationship between the required safety level, the parameters of probabilistic distributions of mechanical loads and resistance to them (strength) and safety factors, as well as the values of the latter, used, for example, in the design of nuclear power plants, can be found in the author's book Spirochkin (2019). The same methodology is applicable in principle to the human tolerance of dangerous effects. The variation of biological parameters is, as a rule, much greater than the variation of mechanical properties, very low in industrial production conditions. Due to the higher variation of biological parameters, it can be expected that the values of $f_k^{(B)}$ corresponding to acceptable safety levels should significantly exceed the values of analogous factors established for structural elements (related to the set of criteria A in Fig. 3.5).

4.2.4 *Minimization of Human Involvement in the Operation of Technical Systems*

Insufficient reliability of technical systems personnel, which cannot be overcome in principle, predetermines a shift in the design paradigm toward excluding humans from management in critical situations and performing operator functions using "smart" automatic controllers (see Sect. 4.2.2). Of course, the intellectual capabilities of these controllers must be perfect (both in the field of processing non-digital information and in computing algorithms), and the faultlessness of their hardware and software components must be exceptionally high. If the system intended to operate in an environment unsuitable for human life (for example, underwater or in space), an additional incentive to minimize human participation in its operation is the complexity of life support. For systems that pose a potential danger due to energy saturation or the presence of harmful technological substances, existing design approaches do not provide guaranteed survival in case of accidents, since due to the large stochastic variations in parameters characterizing dangerous effects on humans and their tolerance, safety levels are not defined unambiguously (see Sect. 4.2.3). Given these circumstances, minimizing the volume of human–machine interaction should be considered actual. The principle of minimizing human involvement in the operation has long been used in the design of technical systems. Currently, this principle is being implemented in a number of directions with the following objectives:

(1) complete exclusion of human operator from the control loop of a technical system;

(2) remote human–machine interaction;
(3) complete exclusion of humans from the operation stage.

The choice of a certain direction is influenced by the functional purpose of the system, the risks associated with its operation in an environment characterized by a perceived degree of unpredictability, and the available technical capabilities to manage these risks. Since technologies are constantly evolving, the boundaries between different directions are not precise, and they have to be determined with some conditionality.

Obviously, the first direction largely corresponds to unmanned aerial vehicles, unmanned cars and trains. Civilian UAVs, which are currently in use, are designed to fly at low altitudes along established routes; they have relatively small dimensions and masses, mainly up to 25 kg (ICAO 2017). Unmanned ground vehicles operate in fairly predictable conditions (known terrain, typical urban road conditions, driving along a certain route with limited speed); therefore, the risks associated with possible accidents are considered acceptable. Human can be in vehicles of this type during their operation only as a passenger. The current level of development of sensors or actuators provides automatic control for normal operating conditions and expected critical situations.

The second direction (remote human interaction with a technical system) is implemented in space, air, water, and underwater vehicles designed to solve complex problems in insufficiently defined operating conditions. Such vehicles are, as a rule, of high cost, and to minimize the risks associated with their loss or non-fulfillment of operation tasks, a combination of automatic control with human analysis of the situation, and work in "director mode" is required, i.e. according to the commands of the operator who gives them remotely. This control principle is used, in particular, in interplanetary space stations and planetary rovers, as well as in remotely piloted aviation systems (ICAO 2015). The latter are created, for example, to solve military tasks related to reconnaissance or strikes. In size, such aviation systems are larger than civilian UAVs, their mass reaches several tons, and they are capable of carrying weapons and spending more than a day in the air. Information about the design, current applications and possible future of various types of UAVs can be found in the book (Austin 2010). The history of the development of military UAVs, the features of their design and operation are described in the publications Holder (2001), Illeez (2015). For all UAVs, the slang designation "drones" is now used; it is also applied to unmanned water and underwater vehicles.

Efforts are currently being made in the world to integrate the traffic control of civilian UAVs, carried out automatically, without human intervention (methods and means of such control are covered by the term "Unmanned Aircraft System Traffic Management", UTM) into national airspace systems. In the United States, the relevant federal agencies and industry have begun to develop a concept of operation, establish requirements for data exchange and infrastructure in order to ensure the operations of these vehicles at low altitudes (less than 120 m above ground level), i.e. in the area of airspace not controlled by the FAA (2020), (NASA 2021). In Europe, Airbus is developing UTM technologies, including those related to unmanned aerial taxis (Airbus 2020). The issues of UAV traffic control, its integration into the airspace and global compatibility are in the field of view of a number of international organizations: in addition to the ICAO, these are the Global UTM Association (GUTMA)—see, e.g., (GUTMA 2017), the European Union Aviation Safety Agency (EASA)—see (EASA 2020), etc.

The complete exclusion of humans from the operation stage of a technical system is possible if its operation takes place in stable, predictable conditions, the reliability of its elements is high enough, and the operations performed by the system are clearly defined (routine) and allow automation in their entire range. This direction, the third in the above list, corresponds to the concept of "unmanned production" (Timofeev and Zlobin 2015; Askarov and Bakytzhan 2017; Malikova 2017). The main limitation in this context is the risk associated with the uncontrolled release of energy accumulated in the system and (or) hazardous technological substances in the event of accident. Given this risk, even if characterized by a low probability, potentially hazardous production systems should be placed outside the human habitat. For nuclear power plants, this direction is implemented with minimal risk, for example, when they are located underground (Dolgov 2006; Bogdanov and Timchenko 2013). Currently, there are seven underground nuclear facilities in operation in the world (they are not completely unmanned production systems); one of these facilities is located in Russia, in *Zheleznogorsk*. Some big businessmen of our time and visionaries have ideas for the transfer of automatically functioning production facilities outside the Earth (Fernholz 2018).

4.3 Managing Uncertainty and Complexity in Design Activities

4.3.1 Accounting for Rare Adverse Events

Rare adverse events accompanied by high mechanical loads and other dangerous effects, for which there are no reliable statistics and (or) reliable estimates of their intensity, except for the assumption of an extreme nature, are a serious design problem (it was already mentioned in Sect. 4.2.2). This problem is related to the interpretation by designers of risks characterized by significant uncertainty, and their consideration in the design with minimal errors.

According to the definition given in the Federal Law of the Russian Federation *On Technical Regulation*, risk is the probability of harm to the life or health of people, property and the environment, taking into account the severity of this harm. The Russian national standard (GOST R 51897-2021) contains terms and definitions related to risk management; it is identical to the international ISO Guide 73:2009. The latter interprets risk as "effect of uncertainty on objectives" with the following note: "Risk is often expressed in terms of a combination of the consequences of an event ... and the associated likelihood ... of occurrence". Regardless of differences in

Table 4.1 Categories and probabilities of events accepted in aviation (AP-25 2009)

Event category	Event probability related to one hour of flight
Probable events, including:	More than 10^{-5}
– Frequent events	More than 10^{-3}
– Moderate probable events	From 10^{-3} to 10^{-5}
Rare events, including:	From 10^{-5} to 10^{-9}
– Low probability events	From 10^{-5} to 10^{-7}
– Extremely unlikely events	From 10^{-7} to 10^{-9}
Highly improbable events	Less than 10^{-9}

definitions, the formal representation of risk for an adverse event is two-dimensional, i.e. it contains two components:

$$\mathcal{R} = \{P, H\} \tag{4.15}$$

where P is the probability of harm as a result of the event;

H characterizes the severity of the harm.

The two-dimensional nature of risk fundamentally complicates the understanding and practical use of this concept. Also, rejection of many specialists is caused by a quantitative assessment of uncertainty using probability. Such an assessment can be very subjective and unreliable, especially if the probability is a priori. In order to unify the understanding of the probabilities of various events and reduce the subjectivity of their a priori estimates, certain categories are used, as it is presented, for example, in Table 4.1 for aviation events.

Based on modern views on safety, when creating technical systems, no potential hazards that can result in an accident or catastrophe can be excluded from consideration. However, the uncertainty of the occurrence of those that have a low probability or are extremely unlikely, as well as the parameters of the corresponding loads and other adverse effects, makes it very difficult to find design and construction solutions that can minimize the risk. Above, in Sect. 4.2.2, some approaches were presented to take into account difficult-to-predict adverse events caused by negative manifestations of human factors (erroneous and malicious actions of the operator of a technical system or his inaction, as well as terrorist attacks). Such events are only part of the entire spectrum of potential hazards: it also includes natural disasters and technology-induced catastrophes, which rather prevail in it. Some of them, for example, hurricanes or earthquakes have no physical limits on intensity. The probabilities of their extreme levels associated with the "tails" of the distributions used by designers cannot be considered sufficiently substantiated. The similarity in uncertainty and difficulties of probabilistic modeling makes it possible to extend the approaches described in Sect. 4.2.2 to all rare adverse events.

The most effective uncertainty management can be achieved by shifting the traditional design paradigm (based on an a priori model of future operation) toward a flexible response to adverse events (regardless of their probability) in accordance with the concepts of Safety II and Resilience engineering. Meanwhile, until such concepts are sufficiently developed for a wide range of engineering applications, there is space for less radical approaches, providing, for example, refinement of probabilistic estimates of rare extreme events based on asymmetric distributions, order statistics and Bayes' theorem.

For example, to describe the parameters of earthquakes, the power law distribution is applicable (Newman 2005). It is possible that such the distribution is also suitable for approximating other natural as well as technology-induced adverse events characterized by the uncertainty of average values, the absence of finite variance and the inverse interdependence of frequency and intensity. But, the substantiation of the correctness of using the power law distribution as a probabilistic model for certain events can create additional (compared to other types of distributions) difficulties (Hilbert 2014).

Probabilistic methods based on the theory of order statistics allow predicting extreme values of random event parameters based on incomplete statistical data. To apply these methods, it is not necessary to know the type of distribution and it is possible to rely on a small number of observations; only the continuity of parameters and the belonging of phenomena to one general totality is required[6] (Gumbel 1962). However, not all natural and man-made adverse events can be considered continuous-parameterizable. For example, an airplane crash at a nuclear power plant, hidden design errors, a violation of the integrity of a structure as a result of through germination of undetected defects during operation are discrete and, rather, cliff-edge phenomena. Extreme values of parameters of such natural events as hurricanes or tsunamis are influenced to a certain extent by the openness of the system, which includes an environment with not fully known external links. As a consequence, the belonging of the events that seem to be of the same type to one general totality cannot be strictly proved. At all, for open systems, especially those whose evolution has not ended, no statistical series, apparently, can be considered a general (in the sense of "complete") totality. Thus, the use of order statistics to assess the risks of rare adverse events should be considered only as a compromise between practical usefulness and mathematical validity.

The possibility to refine a priori Bayesian probabilities exists for events of any origin, but is limited to cases of repetition of the same type of situations belonging to the same statistical totality.[7] For extremely unlikely events, obtaining additional relevant data necessary to refine an a priori estimation (using the Bayesian approach) is a problem difficult to solve. An example of this type of event is the fall of a natural celestial body to Earth: an asteroid or comet. The probability of such events on the territory of Russia is estimated (using astronomy data and methods of celestial mechanics) in the range from 1×10^{-8} to 2×10^{-8} 1/year (Dynamika 1992). At

[6] General totality (or population) is formed by a set of all imaginable qualitatively homogeneous events of the type under consideration, which have the properties of mass character and variability of parameter values.

[7] Unlike the general totality, the statistical totality does not include all imaginable, but only qualitatively homogeneous observed events.

the same time, there is no documentary information about events related to the falls of natural cosmic bodies on nuclear power plants or other energy and industrial facilities. Judging by geological data, large celestial bodies have fallen to Earth in the past. Based on the traces of such events, it is expected that the result of the next collision of a large asteroid or comet with the Earth's surface may be a planetary catastrophe. The a priori probability estimates given above can hardly be refined on the basis of new statistical information. The generally accepted judgment about the exceptional rarity of these events leads to the fact that, despite the extremely high expected losses, no practical measures are taken in the world to minimize the risk.

For comparison: the probability of an airplane crash on the territory of Russia ranges from 2×10^{-6} to 1.5×10^{-5} 1/year, and for an artificial Earth satellite, such probability is estimated as 1×10^{-7} 1/year (Dynamika 1992). These events are classified as low-probability ones (see Table 4.1). There has not yet been a single case of an airplane or other flying object of artificial origin crashing into a nuclear power plant in the world, but the corresponding technology-induced event is included in the design basis of any nuclear power plant being created. Methods were developed for calculating the structural strength of the nuclear reactor containment and other plant buildings based on the mass, kinematic parameters of the falling body, assumptions about the contact area, etc. In order to minimize the risk of release of radioactive substances into the environment due to the destruction of the reactor or spent nuclear fuel storage in such a fall, expensive measures for constructive and other protection are taken.

It can be concluded that the perception of risks of rare events is determined mainly by probability values, and the boundary of differences in their acceptability lies somewhere between 1×10^{-7} and 1×10^{-8} 1/year. Adverse events with an estimated probability of more than 1×10^{-7} are perceived by designers as threats that cause the need for protection, but if the probability is 1×10^{-8} or less, then such events are considered as those that do not require engineering measures to minimize risk, despite the incomparably greater value of losses.

It should be noted that such small probability values cannot be verified practically due to the limited volume of human experience. Indeed, a probability 1×10^{-7} means one event in 10 million years, and 1×10^{-8} corresponds to one event in 100 million years.

In February 2022 (when these lines are being written), the total operation life of nuclear power plants in the world reached 19,219 reactor years (IAEA 2022). During this period, three serious accidents occurred with the melting of reactor cores: at the *Three Mile Island* NPP (1979), the Chernobyl NPP (1986) and the *Fukushima Daiichi* NPP (2011); the latter two because of fatalities or the large losses should be considered disasters. Based on these data, the actual frequency of severe accidents (catastrophes) for nuclear power plants can be estimated as $P = 3/19219 \approx 1.56 \times 10^{-4}$ (1/year). The obtained value characterizes the limitation inherent in human experience in the field of nuclear energy: this experience lasts a little more than 67 years.

The above considerations also demonstrate the possibility of errors when clarifying probability values based on additional statistical data. It would seem that the obtained actual frequency of severe accidents at nuclear power plants can be used to correct the corresponding probability specified in regulatory documents. At the same time, it should be noted that the actual frequency significantly exceeds the values of the probability used in the design—see, for example, Russian regulatory document (NP-001-15): 10^{-5} 1/year for severe accident (accompanied by the destruction of the reactor), and 10^{-7} 1/year for a large radioactive release into the environment. However, the small amount of available statistical data and differences in the factors determining the causes and course of each severe accident hardly provide the necessary reliability of the assessment. Indeed, despite the continuous increase in the number of reactor-years (expansion of the statistical base), the specifics of each of the mentioned events are so significant that they lack the property of homogeneity and it is doubtful that all together they form a statistical totality. In addition, each accident or catastrophe leads to improvements in safety management methods and tools at all operating nuclear power plants, and these improvements as a result change the profile of hazard factors (possible distribution in the space of arguments). Adverse events separated by significant time intervals, during which large-scale corrective measures were carried out to eliminate their causes, cannot be considered to belong to the same type and qualitatively homogeneous.

The question how small the probability of an event should be so that it can be considered practically impossible, and the risk acceptable and not requiring minimization, is beyond the scope of design methods. The interpretation of risk, as well as possible errors in its accounting, are influenced by human factors and, moreover, in such a part of their spectrum that the problem of rare adverse events should be solved not so much at the level of design engineers, but in the plane of design activity management.

Understanding risk as the two-dimensional quantity and maintaining the approach to uncertainty using the concept of probability determines the application of methods that combine two risk components into one composite indicator, for example, in the form of a product (NP-084-15):

$$\mathcal{R} = PH. \tag{4.16}$$

The shortcomings of such the indicator are its dimensionality (which is expressed in units of H and makes it difficult to compare consequences that are different by nature), and difficult-to-control errors in risk assessment due to inaccurate data on the probability of losses and their severity (these data are in most cases very approximate). The elimination of these shortcomings is possible, for example, by ranking the components.

This method is used in the risk matrices (GOST R 51901.1-2002), (GOST R 51901.23-2012). Such matrices allow us to assess the risk of any adverse event, considering the rank of each of the components in the accepted dimensionless scale and assigning appropriate indices to their combinations; based on these indices, we can judge the acceptability of risk. Table 4.2 presents a risk matrix corresponding to the example from the document (ICAO 2018); it is applicable for assessing the acceptability of risk in civil aircraft flights.

Table 4.2 Risk matrix for civil aircraft flights. Based on data from (ICAO 2018)

Probability of adverse event: rank (characteristic)	Severity of losses: rank (characteristic)				
	A (catastrophic)	B (hazardous)	C (major)	D (minor)	E (negligible)
5 (frequent)	**5A**	**5B**	**5C**	**5D**	**5E**
4 (occasional)	**4A**	**4B**	**4C**	**4D**	**4E**
3 (remote)	**3A**	**3B**	**3C**	**3D**	**3E**
2 (improbable)	**2A**	**2B**	**2C**	**2D**	**2E**
1 (extremely improbable)	**1A**	**1B**	**1C**	**1D**	**1E**

The degree of the risk acceptability (based on the tolerability of the expected losses) is determined by the indexes of the matrix elements, which range from 1A to 5E. The risk associated with the indexes 3A, 4A, 4B, 5A, 5B, 5C (matrix elements shaded in red) is considered unacceptable (or intolerable in ICAO terminology). If adverse events considered as possible are characterized by such indexes, it is necessary to immediately take measures to minimize the risk, in particular, to reduce it to an acceptable level. The risk characterized by the index 1A, 2A, 2B, 2C, 3B, 3C, 3D, 4C, 4D, 4E, 5D, 5E (matrix elements shaded in yellow) is considered tolerable. The corresponding levels of risk can be achieved, for example, based on risk-minimizing measures. To accept such a risk in the design, certain management decisions may be required. Risk values indexed as 1B, 1C, 1D, 1E, 2D, 2E (green matrix elements) are acceptable. Measures to minimize them are not required.

It should be noted that the ranking of risk components is not strictly substantiated; it depends on the task under consideration and is carried out for practical reasons. There are no numerical values of probability in the ICAO scale. It can be seen that this scale is expressed only in verbal terms and these terms do not quite coincide with those presented in Table 4.1 (although both examples relate to civil aviation). Only designations of the scale limits in both tables correspond to each other. The possibility of arbitrary ranking can lead to different risk estimates for the same case, to differences in its interpretation and an error in the assessment of tolerability. That is why the problem of rare adverse events should be solved, at least at the current level of knowledge, within the framework of design activity management.

It can also be noted that the boundaries between the areas of unacceptable and tolerable risk (red and yellow areas), or tolerable and acceptable risk (yellow and green areas), if presented on an appropriate scale and in a smoothed form, are similar to curves corresponding to the power law distribution—see Eq. (4.12) and Fig. 4.1. While maintaining the probabilistic approach, it would obviously be useful to carry out a study aimed at mathematically substantiating the assumed inverse relationship between the values of the probability of adverse events and the severity of losses caused by them.

The possibilities of risk accounting are not limited to the methods described above, which reduce the risk to one combined indicator or present it in the form of a matrix. Risk can also be considered as a complex or vector quantity, and appropriate mathematical methods of analyzing such quantities can be used in its assessment.

Another way of accounting for risk—without using the concept of probability and, therefore, being an alternative to the methods discussed above—is the so-called "postulation" of adverse events associated with losses. This method is widely applied in the branches of technology important for safety. Postulation means that the technical system under design must necessarily be analyzed for loads and other adverse effects acting during the supposed event, regardless of its estimated probability. The shortcomings of this method are potential errors in the formation of a list of initiating events to be considered, their combinations with each other and with other dangerous phenomena (see the Swiss cheese model discussed in Sect. 2.2.4), as well as errors in determining the intensity of effects that should be taken into account in analyzes.

To cope with these shortcomings, designers must be able to imagine the worst consequences of adverse events: the types and extent of destruction of elements of the created technical system, human injuries or fatalities, and release of the accumulated energy and dangerous substances into the environment. Such views can become the basis in the design stage for the implementation of constructive measures aimed at ensuring relative safety in the case of accident, and planning organizational measures to minimize losses associated with the alleged destruction. Postulating the most severe damage due to loads or other adverse effects, unpredictable in intensity and place of application, allows solving the problem of rare adverse events, but, obviously, can lead in some cases to excessively conservative design and construction solutions.

A more effective approach combines the postulation of all initiating events that are possible from a rational point of view, with the simulation of scenarios for the development of such events. Within the framework of simulation, the most dangerous combinations of various factors should be considered. This approach seems to be most applicable in the final stages of design that correspond to sufficient certainty in the structural details of the technical system (embodiment design, detail design).

The implementation of all the changes mentioned above in the methods of risk assessment associated with uncertain rare events and ways to compensate for such risks in the design with the lowest possible level of errors belongs to the field of design activity management. Due to the influence of many non-obvious factors, the implementation process itself must be carried out systematically and accompanied by on-site analysis. The proposed changes will contribute to the development of a higher safety culture in the design organization and, as a result, the development of safer technical systems.

4.3.2 Controlling the Complexity of Mathematical Modeling

Design as an area of human activity is characterized by its inherent fundamental complexity, determined, among other factors, by the lack of reliable knowledge about the technical systems being created, especially those that differ significantly from prototypes (see Sect. 3.1.4). Additional aspects of this kind are methodological and computational complexity (Sect. 3.1.2). Methodological complexity is due to the inverse type and fuzziness of the design task (search for a general construction and parameters of a system that should work in specified operating conditions and meet the requirements of reliability, efficiency, and safety). Computational complexity is present both in solving this problem by minimizing the objective function of the design (Sect. 4.2.1), and in various direct problems that arise when substantiating design parameters and planned emergency measures by analyzes (such problems are solved by methods of mathematical physics and operations research). Traditionally, complexity management in the whole is achieved by dividing the design process into stages, during which a gradual detailing of the construction and functional composition of the system is carried out, the searched parameters and planned measures are refined, as well as experimental verification of design and construction solutions is performed (Sect. 3.1.2).

In modern design, there is a tendency of increasing methodological and computational complexity. This tendency is facilitated by expansion of the mathematical modeling area to problems whose solutions previously, due to the insufficient development of computational methods and computer software, were being found not by calculations, but by physical experiments. These problems are related to nonlinear structural dynamics, fluid and gas dynamics, contact interaction, etc. Currently, sophisticated numerical models (mainly based on FEM) are being developed to solve them; typical model can describe in detail the corresponding physical processes using a large number of degrees of freedom. In Sect. 3.2.5, the influence of the number of degrees of freedom (expressed by the number of finite elements) on the modeling error in the simulations of vehicle crashes was discussed. The graph approximating this influence (see Fig. 3.6) looks like a saturation curve: with an increase in the number of elements to 150–200 thousand, the error decreases rapidly, reaching the level of 20%, further growth occurs at a slower rate, and this rate continuously lowers. Models with the number of elements of about a million give the error of at least 10%, while the time needed for calculations performed by the most advanced computer tools reaches 1–2 days.

Such an asymptotic "saturation" in the sense of accuracy (never reaching the ideal) with a simultaneous increase in the cost of obtaining results seems to be characteristic of the numerical solution of any problems of mathematical physics and operations research arising during design. The capabilities of modern software tools in relation to the automatic generation of three-dimensional mesh models covering structures, solid, liquid or gas media and humans during interaction with any degree of detail, as well as simulation of their behavior in a wide range of operating conditions create the illusion of obtaining comprehensive knowledge about the functioning of the designed system. This illusion encourages new generations of designers, who have been accustomed to finite element codes since their student days, to apply them to all emerging problems (including fairly simple ones which have analytical solutions) in any stage of design, and also to hope that numerical modeling and simulation can replace research physical experiments.

However, numerical simulation, due to its algorithmic nature, does not provide demonstrable relationships between design parameters, operational factors and established requirements. Complex mathematical models are not applicable from the point of view of methodology and computational costs in the initial design stages, when unambiguous, well-understandable relationships of this kind and quick verification of decisions being made are required (see Sect. 3.2.5). The capabilities of sophisticated computer tools turn out to be, as noted in Sect. 4.1.3, "traps" for designers, since they provide not knowledge, but data (calculation results), besides containing a significant percentage of errors ("noise"). Knowledge is produced by humans on the basis of practice including physical experience. Mathematical models can be tools of "field research" when factors vary in a certain range, and they can find correlational (but not causal) relationships in the mass of results if human logic is introduced into the research. It cannot be excluded that the production of knowledge from the results of numerical modeling and simulation will be possible when using advanced artificial intelligence—this issue is discussed in Sect. 5.2.4.

The numerical models constructed by FEM cause a specific computational complexity that manifests itself in the simulation process. This complexity is due to the "stiffness" of the corresponding systems of differential equations of dynamic equilibrium. The essence of the stiffness is that the numerical solution of these equations in the time domain includes, besides the main components that slowly change over time, additional "fast" components (Rakitsky et al. 1979; Dekker and Verwer 1984). Such a range of components arises when the characteristic matrix of a system of dynamic equilibrium equations contains eigenvalues that differ greatly from each other. The measure of stiffness can be the spectral condition number of this matrix, determined by the ratio of the modules of the maximum and minimum eigenvalues (Rakitsky et al. 1979). A stable solution of dynamic equilibrium equations can be obtained by an explicit step-by-step integration method only when the time step corresponds to the smallest time constant related to the "fastest" component of the dynamic response spectrum associated with the maximum eigenvalue. Calculating this component and the whole set of similar high-frequency components at each step consumes computational resources. To get a solution in the entire time interval under consideration, a large number of small steps is required. However, such a waste of resources is practically meaningless, since the structural dynamics is determined mainly by the "slow" components of the solution associated with smaller eigenvalues.

The spectral condition number tends to increase with an increase in the number of degrees of freedom of the numerical model. Accordingly, detailed finite element models lead to greater stiffness of the systems of dynamic equilibrium equations and greater complexity of obtaining solution by simulation.

Stiff systems of dynamic equilibrium equations appear not only because of the use of numerical modeling methods. Stiffness originates from the very nature of physical objects, which include parts with different material properties and, accordingly, different time characteristics of the processes inherent in their structure. When such an object is split into finite elements (is meshed) with a high degree of detail, the "natural" stiffness is combined by the "numerical" one, and the resulting stiffness of dynamic equilibrium equations increases. Due to the large number of small time steps required to obtain solution, and the presence in this solution of high-frequency components which have no practical significance and are mainly a consequence of numerical modeling, i.e., a kind of "noise", explicit integration methods are rather inefficient.

This inefficiency can be illustrated by the examples described in Sect. 2.4.4. Here we look at one of them again. This example covers two simulations of the vertical drop test of the Boeing 737 fuselage section. The simulations were performed by two groups of researchers which used two finite element models similar in structure: the model of the first group included 13,600 elements, and the model of the second one included 56,700 elements (the latter was generated based on the 3D model shown in Fig. 2.13). Parameters of structural dynamics were calculated by two explicit nonlinear program codes: MSC.Dytran (Jackson and Fasanella 2001) and LS-DYNA (Byar et al. 2001). In both simulations, the calculated acceleration time histories were filtered to "clean up" from high-frequency oscillations (corresponding to the "fast" components of the solution, considered mainly as "noise" due to errors in numerical modeling). Despite a very small integration step (~10^{-6} s), only 2–3 oscillatory components with frequencies approximately from 25 to 50 Hz were present in the obtained time histories.

In Sects. 2.4.6 and 2.4.7, examples of simulations of car accidents and rail vehicle crashes are given, which are characterized by similar contradictions between the time step when solving the dynamics equations and the oscillation modes present in the dynamic response—e.g., see Fig. 2.28. Herewith, in the course of calculations performed by explicit methods, it became necessary to control and limit non-physical effects caused by significant changes in the geometry of elements due to large displacements. Modern attempts to exclude these effects by means of meshfree approximations lead to an additional increase in calculation time.

There are more efficient integration methods that allow solving stiff differential equations with time steps of the same order as large time constants associated with small eigenvalues (which correspond to the main components of dynamic response). These include, for example, the implicit methods of Hubolt, Newmark and Wilson (Bathe and Wilson 1976). They were used in simulations of emergency landing of the *Buran* spacecraft (see Sect. 2.4.3) and in solving other similar problems of nonlinear dynamics (Lyakhovenko et al. 1989; Spirochkin and Shenk 1990; Spirochkin 1993).

To solve stiff dynamic equilibrium equations during a simulation, other numerical integration methods can be used, for example, modal analysis methods based on the Fourier transform (they are mentioned below). But the applicability of such methods is mainly limited to linear dynamics problems.

In general, the "protection" against excessive complexity of mathematical models, accompanied by insufficient accuracy of simulation results and leading to inefficient implementation of the "design by analysis" approach, should be based on simplification of modeling. One of the ways to reduce complexity, minimize errors in the interpretation of "noisy" simulation output and increase the efficiency of "design by analysis" according to the criterion "accuracy/cost" is regulation (Silver 2012), with reference to the book *Future Shock* by A. Toffler. Such regulation relates to the field of design activity management; it should be applied at all levels of management and include standardization measures.

To date, standards have been developed on various aspects of numerical mathematical modeling and simulation; among them:

- NASA's technical standard (NASA-STD-7009A 2016);
- documents ASME V&V 10-2006 (Guide for Verification and Validation in Computational Solid Mechanics), ASME V&V 10.1-2012 (An Illustration of the Concepts of Verification and Validation in Computational Solid Mechanics), ASME V&V 20-2009 (Standard for Verification and Validation in Computational Fluid Dynamics and Heat Transfer);
- international standards ISO 19364:2016 (Passenger cars—Vehicle dynamic simulation and validation—Steady-state circular driving behavior), ISO 19365:2016 (Passenger cars—Validation of vehicle dynamic simulation—Sine with dwell stability control testing), ISO 16730-1:2015 (Fire safety engineering—Procedures and requirements for verification and validation of calculation methods—Part 1: General), ISO/TR 16730-2:2013 (Fire safety engineering—Assessment, verification and validation of calculation methods—Part 2: Example of a fire zone model), ISO/TR 16730-3:2013 (Fire safety engineering – Assessment, verification and validation of calculation methods – Part 3: Example of a CFD model), ISO/TR 16730–4:2013 (Fire safety engineering—Assessment, verification and validation of calculation methods—Part 4: Example of a structural model), and similar documents;
- Russian standards (GOST R 57188-2016), (GOST R 57700.1-2017), (GOST R 57700.2-2017), (GOST R 57700.4-2017), (GOST R 57700.6-2017), (GOST R 57700.7-2017) and other ones of this series.

None of the listed documents contains provisions regulating the complexity of mathematical models; they mainly consider ensuring the accuracy of numerical calculations.

Modeling issues are addressed in regulatory documents used in nuclear industry. For example, Russian Federal regulations and rules (FNP) do not directly regulate the calculation methods and mathematical models applied in the analysis of the strength and safety of nuclear facility components. The regulatory approach covers the establishment of

- criteria for the strength and integrity of the structure to be checked in the form of the expressions (3.1);
- measures of strength and integrity $\mathcal{S}_k$ for each k-th limit state ($k = 1, 2, \ldots K$), for example, allowable stresses $[\sigma]$;
- requirements for determining the loading measure $\mathcal{F}_k$ related to this limit state and expressed in terms of the stress–strain state, e.g., in the components of the stress tensor $\{\sigma\}$.

The choice of the calculation method for determining the components of $\{\sigma\}$ is left to the analysis engineer. In the Russian regulatory document (PNAE G-7-002-86), some recommended methods of strength analysis are presented, in which numerical models with lumped (concentrated) elastic and inertial parameters are used. However, they are included in appendices that do not have a mandatory status. To determine the dimensions of structural elements, which are carried out in the initial stages of design, this regulatory document provides for analytical and semi-empirical dependencies—in fact, simplified mathematical models, and their use is mandatory.

The same approach is typical for nuclear energy regulatory systems in the USA, Western Europe and other regions of the world. The description of modern numerical methods can be found only in non-mandatory parts of a few regulatory documents, for example:

- Appendix N: Dynamic Analysis Methods, in the ASME Boiler & Pressure Vessel Code (ASME 2019), Section III, Division 1;
- Annex C: Calculation Methods, in the KTA Safety Standard (KTA 3201.2 2017).

It is noteworthy that these materials do not contain clear rules for constructing mathematical models that would ensure effective complexity management. The recommendations given in this regard are too general or vague. It seems that the very existence of modern computer programs of CAE type, equipped with the functions of automatic creation of numerical models of any level of complexity and formal control of some aspects of their accuracy, eliminates the need for general logical control of modeling by humans. However, to prevent excessive complexity of numerical models, improve simulation efficiency and minimize "noise" in the results, well-defined and practical requirements and recommendations are needed. It would be advisable to include in regulatory documents, for example, the following provisions:

1. Mathematical models used in the design must be compatible with the traditional design technology, which provides for the division of the design process into stages associated with the gradual detailing of the created technical system, step-wise elaboration of design and construction solutions and, if necessary, their experimental verification.
2. The complexity of mathematical models should correspond to the purpose of calculation (simulation); this purpose is determined by the design stage and the accuracy of available input data. All other things being equal, preference should be given to models with a minimum level of complexity.
3. Differences in the content of the work performed in different design stages, and in the depth of elaboration of design and construction solutions determine the development of a number of models of several levels of complexity (from analytical to numerical ones with different numbers of degrees of freedom), even for solving problems of the same type.
4. In the initial design stage, it is recommended to use models of the system and its elements in the form of analytical functions that clearly represent the relationship between the basic design parameters of the system, operational factors and established requirements.

5. Models of different levels of complexity should provide the possibility of mutual transformation with compliance on the basic design parameters: overall dimensions, inertial characteristics and interfaces (connections of elements to each other and the interaction of the system with the environment).
6. Each of these models should allow verification and correction, including that on the basis of experimental data.
7. In each of the design stages, the traceability of the input data for modeling, as well as the verifiability of the simulation results, should be ensured.

Traditional design technology is, in fact, an invariant that that exists for a long period of time (approximately corresponding to the XX century). During this period, such major technological achievements as the creation of aviation, nuclear energy and the rocket and space industry were made. The emergence and development of computers, advances in numerical modeling and software tools provided, in fact, only the improvement of previously found fundamental technical solutions, but not the discovery of new and breakthrough ones. Numerical mathematical models are subject to obsolescence, in this respect, they are similar to hardware or software updated with a cycle of several years (as discussed in Sect. 5.1.2). When updating, there is a risk of losing the knowledge implemented in them (see Sect. 5.1.3). Compatibility with traditional design technology, which fixes solutions of the relevant problems by the human mind and is invariant in relation to computing tools, can to some extent guarantee protection against loss of knowledge.

An integral part of this technology is the determination of the main parameters of the designed system (dimensions, mass characteristics of elements, etc.) in the initial design stage using simple analytical and semi-empirical models mentioned above.

Such models fully meet the objectives of conceptual design and allow designers to determine input data for the next stage. For example, when designing space systems, input data include distributions of mass and stiffness parameters along the longitudinal axes of the main structural assembly units: launch boosters, second-stage rocket blocks, etc. The unavoidable error in these distributions in the preliminary design stage can be tens of percent. An analysis engineer can work with such low accuracy using models built from beams and springs approximating mechanical behavior of the corresponding assembly units. Beam-spring models allow designers to estimate the low-frequency modes of assembly vibrations and associated dynamic loads, which, as a rule, determinate structural strength. The dynamic characteristics of these modes (natural frequencies and mode shapes) and dynamic loads (maximum values and time profiles) are the input data for further elaboration of the design in detail and substantiation of structural strength, as well as the development of automatic control system. Thus, the whole process consists of iterations that are repeated in the design stages, based on increasingly improving design parameters.

Simple models built from finite elements in the form of beams and springs with refined mechanical properties can be so effective that they are applicable not only in design, but also in support of operation by analysis. A model of this type, developed at NPO *Energia* in 1995, was used to analyze the structural dynamics of Russian space station *Mir* in the case of the *Space Shuttle* docking with it. The analysis was

From a NASA video on February 6, 1995

a

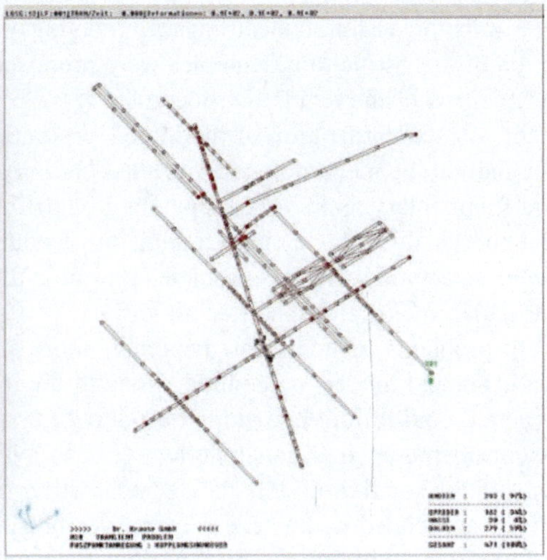

b

Fig. 4.3 **a** Russian space station *Mir* (Reproduced from https://history.nasa.gov/SP-4225), **b** Its finite element model (Reproduced from (ISAFEM3D 2020)). Courtesy of Dr. Gerhard Krause

carried out, of course, before the docking, but after the completion of the design of the station, when it has been in Earth orbit for several years. In Fig. 4.3a, the real configuration of the station is shown, and Fig. 4.3b presents its finite element model created using ISAFEM3D software, a product of Dr. Krause Software GmbH, Germany.[8]

When analyzing the dynamic behavior of a technical system, minimizing the complexity of numerical mathematical models could be achieved by:

- reducing the number of spatial dimensions of the stress–strain state in structural parts or units to a level corresponding to one or two main components of this state (for example, in beams or springs, the stress–strain state can be considered one-dimensional);
- reducing the number of degrees of freedom by dividing them into primary and secondary ones; in the model (and in simulation the former are taken into account directly, and the latter are calculated on the basis of their values using kinematic or other relationships);
- transforming the coupled degrees of freedom to modal (uncoupled) coordinates (Fourier transform) and choosing a minimum subset of these coordinates which corresponds to the main vibration modes, etc.

---

[8] The current name of the company is Dr. Krause GmbH. ISAFEM3D was chosen by Russian analysis engineers because of the compatibility of this program with the *Space Shuttle* finite element model used by American partners.

Similar provisions concerning the modeling of nuclear facility elements and simulation of their structural dynamics were proposed by the author for inclusion into a new edition of the regulatory document (PNAE G-7-002-86), as well as into documents on standardization of the State Corporation Rosatom.[9] These provisions do not claim to be methodological novelty, but only streamline existing heterogeneous and fragmentary approaches depending on the characteristics of elements, external and internal dynamic loadings, taking into account the experience of other industries where structural dynamics problems prevail in the spectrum of tasks to be solved (in particular, rocket and space technology).

In problems of nonlinear dynamics, more subtle control of the complexity of mathematical models is required. None of the methods for solving stiff systems of dynamic equilibrium equations (see above) arising from numerical modeling can be considered unconditionally effective. Explicit integration methods are applicable only with a small time step; they generate high-frequency "noise" and non-physical effects associated with changes in the geometry of finite elements because of their deformations. Implicit methods allow large time steps, but due to the nonlinearity of the resolving system of algebraic equations, iterative algorithms are necessary, and difficult-to-control errors are possible in the solution. The use of modal analysis methods is feasible only in the case of linearization, what is possible at a small degree of nonlinearity. It can be argued that the common reason for the inefficiency of all existing time integration methods is the isochronous principle of modeling and simulation, according to which processes with different time characteristics are taken into account by a single discretization of the time coordinate.

For example, the results of numerical simulation of vehicle crashes presented in Figs. 2.10, 2.18 and 2.28 allow us to conclude that the obtained acceleration histories in all cases have common features:

- in each acceleration-time graph, there is a basic pulse of a sinusoidal (trapezoidal or triangular) shape, the duration of which ranges from several tens to hundreds of milliseconds; its parameters are determined by the general geometry and mechanical properties of the structure, as well as its velocity at impact;
- vibrational components with frequencies of the order of tens Hertz with smaller amplitudes are superimposed on this pulse, they correspond to the dynamic response of some structural elements.

This feature of structural dynamics, in principle, makes it possible to divide the structure into the corresponding areas of predominantly "slow" and predominantly "fast" response. When considering the first area, it is advisable to analyze the general displacements and deformations of the entire structure with a large time step, and in the second area, local deformation processes should be simulated with a much smaller time increment. There is a mechanical interaction between these areas: the "fast" area

[9] These provisions were included in the draft documents prepared by the author in 2019–2021 while working as the chief specialist at the Engineering Center for Strength, the rulemaking department of NIKIET.

of the structure affects the "slow" one through vibrations that undergo physical "low-frequency filtration" due to the greater mass of the latter, and the reverse effect of the "slow" area on the"fast" one manifests itself as a quasi-static load over several cycles of vibrations. The total mechanical reaction can be considered as the sum of its "slow" and "fast" components, meanwhile their superposition, or interaction, can be taken into account periodically, after a certain number of small time steps or each large time step. This method of modeling and dynamic analysis can be called "non-isochronous".

In order to regulate the complexity of mathematical models, it seems useful to standardize the methods of finite element meshing for typical structural parts. Such standardization, based on proven solutions, including those confirmed by experiments, would avoid errors caused by insufficient qualification of analysis engineers.

On the other hand, it would be advisable to direct efforts to approximate the data obtained during mechanical tests and computational simulations by suitable analytical dependencies. For example, the above-mentioned common feature of acceleration pulses in vehicle crashes allows us to use for approximation the following generalized expression:

$$a_{\max} = a_{\max}(b, r, M, f_1, \lambda_l, \sigma_l, V) \tag{4.17}$$

where b is a characteristic dimension of a vehicle, which determines the area of application of the impact load (in the case of aircraft it is the diameter of fuselage);

r is the radius of curvature of the surface of this area (affecting the gradient of the impact load increase during deformation of the vehicle structure along the normal to the surface);

M is the mass of the vehicle;

f_1 is the main (as a rule, the first) frequency of the structural vibrations excited by impact;

λ_l is the local flexibility of the vehicle structure in the area of application of the impact load, along the normal to the surface;

σ_l is the stress that characterizes the local loss of the load-bearing capacity of the vehicle structure in the area of application of the impact load;

V is the velocity of the vehicle at the beginning of the impact, in the direction corresponding to the normal to the surface where the impact load is applied.

In order to transform this generalized expression into the desired parametric dependence, physical experiments and simulations in a wide range of conditions are necessary. As a result, an analytical model or a number of compatible models can be created that will capture the relevant data in the form of knowledge.

Chapter 5
Human Interaction with Long-Lived Technical Systems

5.1 Long Life Cycle Issues

5.1.1 Aging

Long-lived technical systems can be considered those whose service life is comparable to or exceeds the duration of human life. Such systems are, for example, nuclear reactors and nuclear power units as a whole. As of February 2022, there are 439 nuclear reactors operating in the world with a total electrical capacity of 390,624 MW (IAEA 2022). The 296 power units providing 67.1% of this capacity are at least 30-year-old. The share equal to 26.9% is generated by 133 power units, which are in operation 40 years and more. Approximately, 3% of the total electrical capacity belongs to 20 power units, whose age ranges from 50 to 53 years. The five 53-year-old units connected to the grid in 1969 include two reactors in India (*Tarapur*-1 and *Tarapur*-2), one in Switzerland (*Beznau*–1) and two in the USA (*Nine Mile Point*-1 and *Ginna*). The WWER-1200 reactors, by which new Russian nuclear power plants are equipped, must operate for at least 60 years (Rosatom 2020). Taking into account the duration of design and construction work (about 10 years), as well as the decommissioning of the power unit (up to 30 years), the entire life cycle will be about 100 years (IAEA 2020).

The *Voyager* 1 and *Voyager* 2 automatic spaceships were launched 1977. As energy sources, they use radioisotope thermoelectric generators containing plutonium-238 (its half-life is just over 87 years). The power of the generators decreases over time, and accordingly, the functionality of the installed equipment reduces. It has been a long time since they left the Solar System, but they continue to transmit information. In November 2017, after 37 years of downtime, the *Voyager* 1 engines were turned on at the command of a team of NASA specialists to correct orientation in space (Landau 2017). The service life of both *Voyagers* has already reached about 45 years, and it is hoped that communication with them will be able to be maintained until 2025.

Y. Spirochkin, *Human Factors and Design*,
https://doi.org/10.1007/978-981-19-8832-5_5

The International Space Station has been in operation since the end of 1998. The participating countries of the project have agreed to continue it until 2024, and the possibility of extending the service life for another 6 years was considered (RIA Novosti 2020). Then, the duration of operation will be 32 years.

Some airplanes produced in the USSR, such as IL-18 or AN-12, have been flying for more than 50 years. American B-52 bombers, assembled in the late 1940s–early 1950s, are still in operation. Modern air vehicles are designed for a service life of 30–35 years or 60,000–70,000 flight hours (Boiko 2010, Nesterenko 2010).

With such long-term operation, aging of any technical system occurs, caused by the corresponding physical and chemical processes. Aging is expressed in deterioration (gradual adverse change) of mechanical properties and geometric parameters of the structure under the influence of operational factors (mechanical and temperature loads, neutron radiation, corrosion and erosion), as well as in degradation of electrical, electronic and other components. The duration of operation is determined mainly by the exhaustion of structural resistance (strength) to fatigue damage under alternating mechanical and thermal loads in combination with other effects of aging.

The assigned duration of the service life is based on the design model of operation, which covers the expected loading processes and the predicted rate of degradation of components. Due to the conservatism of this model and a priori forecasts of aging, the actual service life usually exceeds the assigned value. Prolongation of the service life can be ensured by monitoring the system health during operation, evaluating the exhausted and residual life, regular maintenance and necessary repairs.

In order to ensure the long-term operation of nuclear power units, appropriate aging management programs are being developed (IAEA 2018). Similar measures (without using the term "aging management") are also implemented to maintain the airworthiness of aircraft—see, for example, Russian *Aviation Rules* (AP-25 2009), § 25.571. In the aviation of developed countries, it is believed that an aircraft ages after 15 years of operation (Nesterenko and Nesterenko 2014). During this period, fatigue and corrosion damages occur and progress in the structure, and characteristics of electrical and electronic components change so that it is necessary to solve the question of the economic efficiency of further operation (taking into account control, maintenance and repair). Approaches to forecasting the aging of aerospace vehicles and structures can be found in the book Ho (2010).

Human factors manifest themselves in determining the assigned service life of long-lived technical systems and managing their aging to ensure safe and efficient operation. Despite some controversial and debatable issues, in industry in general, a fairly reliable technology for solving relevant problems has been developed.

5.1.2 Obsolescence

Along with aging, long-lived technical systems are also subject to obsolescence. This term refers to the discrepancy that occurs over time between the design and construction solutions implemented in the system in question and the current, more developed state of science and technology or new variants of products with which it is connected through life cycle processes (purchased components, tools, etc.). As a result of this discrepancy, the use of the system for its intended purpose becomes difficult, economically unprofitable or even impossible.

Nowadays, examples of obsolescence are aircraft flight control systems which are not hydraulically or electromechanically powered, analog electronic devices, electron beam displays, magnetic tapes, floppy disks, etc. Obsolescence occurs much faster than aging: the average duration of the obsolescence period is estimated as 5–7 years (Il'enkova et al. 2002). Especially, rapid obsolescence associated with frequent updates of components is typical for computer hardware and software.

Replacement of obsolete products and technologies with new ones may be accompanied by incompatibility with the environment, human habits, skills and traditions, loss of information, other negative circumstances or direct costs. Manufacturers and repair organizations usually stop supporting products that are considered obsolete, because it becomes unprofitable to keep production facilities and spare parts in stock with a decrease in the number of customers. This leads to a shortage of spare parts and qualified repair personnel to maintain obsolete but still used products; as a result, the costs of maintaining their life cycle tend to increase.

Obsolescence and continuous change of products, technologies and, in general, methods of solving scientific and technical problems are another manifestation of human factors that must be taken into account when designing systems intended for a long service life. Different approaches are possible in this regard.

One of them is expressed in the answer given by the Russian chief inspector of shipbuilding A. N. Krylov[1] to the question of the Defense Commission of the State Council in 1908: will the battleships being created become obsolete in four years of construction? We can read in his memoirs—see, e.g., the 9th edition (Krylov 2016): "… first of all, it is necessary to give an exact definition of what you mean by the word "obsolete". It is usually required that the ship being developed will be, at the beginning of the design, the strongest ship in the world. If so, then I will answer that our battleships will become obsolete not in four years of their building, but from tomorrow". And to explain this provision, the famous Russian shipbuilder says that the chief inspector of shipbuilding of a political adversary, "starting to design a battleship tomorrow, he will receive the same requirement … and will have to take into account our ship and develop a ship stronger than ours. It is necessary to take care not of a single day, but to anticipate what is possible and design the ship so that it remains combat-ready and powerful for a possible long time".[2]

According to another approach that appeared in the second half of the twentieth century, due to obsolescence, it is not always advisable to design a technical system for too long durability (Komarov 1969).

---

[1] Krylov, Aleksei Nikolaevich (1863–1945) is a Russian and Soviet mathematician, mechanical engineer and shipbuilder, a member of the St. Petersburg Academy of Sciences, then of the USSR Academy of Sciences, Professor of the Naval Academy, Fleet General, a member of the St. Petersburg Mathematical Society, an honorary member of foreign scientific and engineering societies and the founder of the modern Russian school of shipbuilding.

[2] Translated into English by Y. S.

In connection with the development of optimal design methods, the idea of including resistance to obsolescence in the objective function of the design was considered. But there are difficulties in measuring and controlling such obsolescence, since it is caused mainly by unforeseen advances in technology (Hall 1965).

In accordance with the Russian methodology of mechanical engineering, when developing products, two or three upgrades should be provided, while the main load bearing structures, mechanisms and assembly units should be preserved without significant changes (Klyuev 2003).

The modern approach is to manage obsolescence; it is outlined in the European standard EN 62402:2007 (Obsolescence management—Application guide), the corresponding standard of International Electrotechnical Commission IEC 62402:2007 and International Railway Industry Standard, Guideline 5 (IRIS 2012). Some aspects are reflected in the Russian national standards (GOST R 56129-2014) and (GOST R 56136-2014). This approach is implemented within the framework of system life cycle management. The possibility of obsolescence should be considered in the earliest stages of design, and appropriate management should be provided. This management should include design and organizational measures aimed at reducing potential losses from obsolescence, as well as monitoring the manifestations of obsolescence during operation.

The measures to minimize potential losses should be fixed in the obsolescence management plan, which provides (Goring 2016):

- life cycle sustainability in relation to product items, parts, materials and technologies;
- supply strategy (identification of preferred manufacturers of purchased components and the use of multisource items);
- technical strategy (modular construction of the system, transparent and open system architecture);
- identification of critical components (with reduced dependence between hardware and software).

The basis for formation of the plan is an analysis of the risk associated with obsolescence. This analysis and the choice of a management strategy (active or passive) shall be carried out, as stated in the Russian national standard (GOST R 56129-2014), in the stage of technical design, which is defined in Sect. 3.1.2. An active strategy involves the early implementation of measures that reduce the likelihood of negative consequences of obsolescence. According to (GOST R 56129-2014), the obsolescence monitoring is recommended for:

- components, parts and materials with the highest cost;
- the only suppliers;
- rare professional skills;
- supplies of components that provide critical system functions;
- products with long service life.

An effective way to protect against obsolescence is the above-mentioned construction of a technical system from modules that can be replaced; this ensures modernization and the possibility of new applications. The development of unified modules can be an independent process, not directly related to the design of the system. Currently, space technologies and nuclear energy are developing along this path in the world. A promising direction of protection against obsolescence may also be the appropriate adaptation of the Resilience engineering concept, for example, in the form of "obsolescence resilient design" (Goring 2016).

5.1.3 Loss of Design Knowledge

When a technical system is operated for a long time, there is a risk of losing the knowledge on which its design was based. The specialists who participated in the creation of the system grow old and leave, taking with them the accumulated experience and a certain amount of corporate memory. An essential part of their "intellectual baggage" is implicit knowledge that is not recorded in the technical documentation. With many years of skills, and information related to "know-how", professionals are able to answer complex technical questions to which representatives of the new generation have no answers. The loss of old key specialists together with their knowledge poses a potential threat to the safe and reliable operation of long-lived systems. This situation is typical for any high-tech industry. In nuclear power industry, it becomes particularly acute due to the extension of life cycle of nuclear power plants to 100 years (see Sect. 5.1.1), which exceeds the period of labor activity of at least two human generations. The manifestation of human factors in the form of loss of design knowledge should be taken into account when creating a technical system along with its aging (Sect. 5.1.1) and obsolescence (Sect. 5.1.2). In the framework of design, measures should be provided to minimize the potential harm associated with the loss of knowledge on which it was based.

The problem of loss of design knowledge has technical, social and economic aspects. In nuclear power industry, the first of them can be characterized as follows. Of the total number of nuclear power units currently operating in the world (see Sect. 5.1.1), 328 reactors were put into operation before 1990 (IAEA 2020). Most of the information related to their design was originally presented on paper or in databases of various suppliers, based on different goals. Heterogeneity of information storage methods and formats, as well as the technical means used, largely determines the risk of loss of design information. For the safe operation of a nuclear reactor and making effective decisions on maintenance or termination of its operation, it is necessary to ensure the integrity, reliability and availability of this information to all participants in the life cycle.

The social aspect comes to the fore when the development of any industry branch stops. For example, the countries that have refused from nuclear energy and stopped building new nuclear power plants face the following challenges (IAEA 2006b):

(1) retention of existing skills and competencies during several decades for decommissioning nuclear power plants and recycling their elements;
(2) development of new skills and competencies for phased decommissioning and radioactive waste management in conditions when young people no longer seek to study nuclear sciences due to the lack of any prospects for their application in the future.

The difficulty of responding to these challenges is aggravated by the economic aspect of the problem under consideration: due to the competition of nuclear power with other types of energy generation, there is a need to reduce operating costs, reduce personnel and training measures, as well as reduce R&D at the industry level. It is obvious that to solve the whole complex of emerging issues, an adequate technical policy is required. Some issues relating to technical policy are discussed in Sect. 5.2.5. Here, we will pay attention to the technical aspect of the problem and turn to knowledge management technologies that should prevent the loss of design knowledge.

These technologies are aimed at the effective retention of knowledge, including implicit knowledge and ensuring their transfer to successors. Currently, technical systems are created on the basis of three-dimensional modeling in computer programs of the CAD category and calculations performed by software codes of the CAE type, using electronic databases. The developed models, design and construction documents, technological and operational documentation, as well as the regulatory framework for the design, manufacture of elements and construction of the system, and other relevant sources of information form an *information environment* that should become available to the operating organization when the created system is commissioned. Further, this information environment should be supplemented with data characterizing the operation of the system, including information on repairs and replacements of elements, in order to ensure a long service life (with the possibility of extending it), aging management and decommissioning of the system.

The concept of the information environment first appeared in aviation and was being implemented using Continuous Acquisition and Life cycle Support (CALS) technologies. Then, this concept was adapted and refined in other industries. In nuclear power, it is being developed at present within the framework of Plant Life Management (PLiM). In the Russian nuclear sector, the relevant activities were carrying out as the topic of the State Corporation Rosatom "Information support of the product lifecycle", abbreviated IPI[3] (Spirochkin 2019). The team that started the development of this topic[4] considered IPI as a combination of software, hardware and a third component—the so-called *dataware*.

From the point of view of IAEA, the goal of effective retention of design knowledge about a nuclear power plant is achieved by creating and applying its information model (IAEA 2020). This model is a set of information about plant elements and systems, including data, relationships between them and rules for describing objects and processes throughout the life cycle. A plant information model should provide:

[3] From the transcription of Russian term "Informatsionnaya Podderzhka Izdelii".

[4] The author was the head of this topic at NIKIET in 2006–2008.

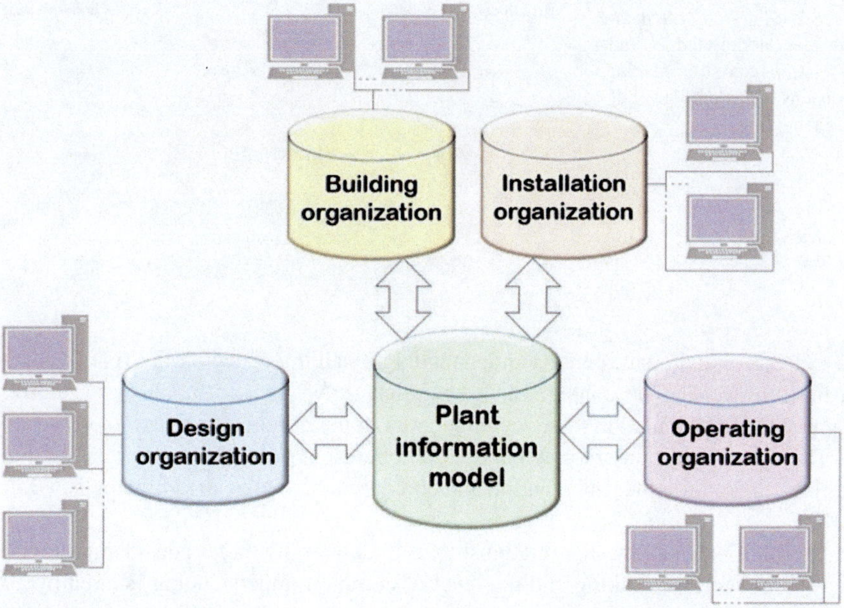

Fig. 5.1 Data exchange between participants in the life cycle of a nuclear power plant based on plant information model

- integrity and reliability of the design basis;
- competence of operational personnel;
- data exchange between participants of the plant life cycle, first of all, the transfer of design knowledge and information from design organizations, building and installation companies to the operating organization (see Fig. 5.1).

The use of such a model should reduce the design implementation time and minimize the impact of human factors by preserving design knowledge in electronic form, using a single source of information and "seamless" data transfer between participants and stages of the life cycle, as well as automated data processing.

However, there is currently no concept of a knowledge-based information model agreed by all stakeholders, and the existing exchange standards are also not generally accepted. Moreover, the terms "data", "information" and "knowledge" are not defined clearly. The precise definition is hindered, apparently, by the fundamental nature of the entities behind these terms. Practitioners consider, instead of definitions, the relationship between data, information and knowledge. These relationships form a hierarchy in the form of a pyramid, at the bottom of which lay data (see Fig. 5.2).

Data contain symbols, figures and facts representing material objects, events or phenomena in an unstructured form. After processing, structuring and linking together in a certain way, data become interpretable and readable, providing an accurate description of the object or phenomenon in question. Thus, data transform into *information*. It turns into *knowledge* in the presence of a context sufficient for

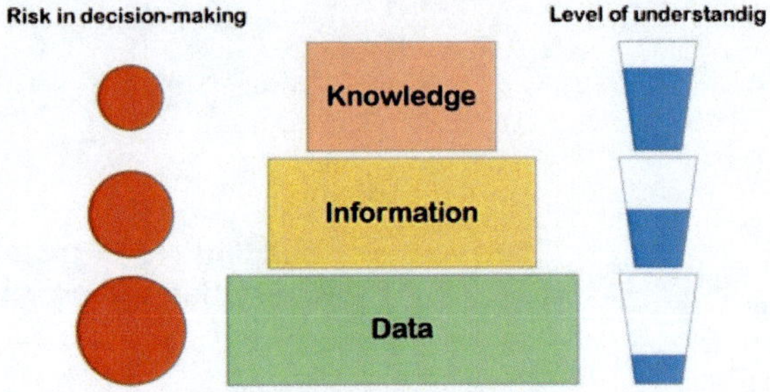

Fig. 5.2 Hierarchy of "data-information-knowledge" and its relations to risk in making decisions and shared level of understanding

an exhaustive and unambiguous understanding by all interested parties (participants in the life cycle). The context can be implicit (unwritten), difficult to formalize, fragmented, distributed in many documents or in the minds of experts.

There is a definition of knowledge as "a conclusion drawn from information that has been linked to other information and compared to what is already known" (Daft 2010).

Figure 5.2 also shows the relationship of the hierarchy under consideration with the risk in decision-making and the level of common understanding (shared by all stakeholders). In the left part of the image, it can be seen that if decisions (made, for example, by an operating organization) are based more on knowledge than on information, they are accompanied by less risk. As the right part of the image shows, the completeness and clarity of description, which usually are associated with knowledge, contribute to a better understanding of a problem by all stakeholders involved in the life cycle.

The concept of information model applicable during the life cycle of a nuclear facility was being developed within the framework of the rulemaking activity of the State Corporation Rosatom. Some proposals to this concept, detailing the relevant provisions on the topic IPI (see above), were proposed by the author during the preparation of new industry draft standards.[5] These proposals concerned the definition, structure and attributes of information models for components of designed nuclear power plants and their life cycle processes.

Information modeling in different industries is developed unevenly. In this respect, the building industry and the area of facility management are characterized by the highest level: there are international standards for describing construction objects throughout the life cycle, documenting data and exchanging them. These standards include:

- ISO 16739-1:2018 Industry Foundation Classes (IFC) for data sharing in the construction and facility management industries—Part 1: Data schema;
- ISO 29481-1:2016 Building information models—Information delivery manual—Part 1: Methodology and format;

[5] The preparation work was carried out in 2017, and the author took part in it as the chief specialist at the Engineering Center for Strength, NIKIET.

- ISO 29481-2:2012 Building information models—Information delivery manual—Part 2: Interaction framework (last reviewed and confirmed in 2018);
- ISO 12006-2:2015 Building construction—Organization of information about construction works—Part 2: Framework for classification;
- ISO 12006-3:2007 Building construction—Organization of information about construction works—Part 3: Framework for object-oriented information (last reviewed and confirmed in 2012).

We shall also mention the relevant Russian national standards (GOST R 10.0.02-2019), (GOST R 10.0.03-2019), (GOST R 10.0.04-2019), (GOST R 10.0.05-2019), (GOST R 10.0.06-2019).

In recent decades, R&D have been carried out in the world aimed at standardizing the exchange of information. They have led to the creation of unified file formats that allow various software codes to read, to process and to transmit existing types of information: e.g., DXF for drawings, STEP for 3D objects, PDF for text documents, etc.

The standards of the ISO 15926 series developed to regulate the exchange of information and the corresponding Russian national standards (GOST R ISO 15926-1-2008) and (GOST R ISO 15926-2-2010) determine not only electronic compatibility, but also semantic aspects and consistency of formats. However, the results of standardization still do not satisfy all the needs of participants in the life cycle of technical systems. In particular, the issues of compatibility for mathematical models of different dimensions (for example, 3D and 2D) intended for different purposes (structural design and engineering analysis) and different levels of complexity, including their mutual transformations, remain unresolved. In addition, neither information modeling nor information exchange standards ensure the fixation and transfer of implicit knowledge. This knowledge can be related to the current paradigm of technological development, and an adequate definition and understanding of such a paradigm are achieved only at the level of philosophy. The latter, as a rule, does not attract the attention of developers and researchers, engaged in solving practical problems, and therefore falls out of the field of formalization.

It should be noted that the concepts described above and the mentioned data formats appeared relatively recently—during the lifetime of one human generation. It is unknown how they will develop further. In Sect. 5.1.2, the possibility of information loss due to incompatibility of new and old generations of products and technologies was discussed. Such incompatibility is very likely for computer hardware and software—and in general, for information technology—due to their rapid obsolescence and frequent updates. All types of information are currently stored mainly in electronic digital form; at the same time, complete independence of coding methods from processing technologies cannot be considered guaranteed. Increasing the amount of stored information complicates its transformation (recording) when changing technologies. This is an additional risk factor.

The solution of the problem associated with the loss of design knowledge cannot be reduced only to the retention and transmission of information or to the reliable fixation of knowledge itself. This problem also has a cognitive dimension. Experts in the field of neuropsychology have noticed that the development of information technology and its widespread use in society are accompanied by the degradation of mental activity. The results of the relevant studies are presented, for example, in the publications (Kurpatov 2018; Kurpatov 2020). Related issues are discussed in Sect. 5.2.4.

Finally, there are doubts that new generations of engineers will be able to adopt the knowledge of their predecessors to the required extent, even if all this knowledge is properly recorded and stored, without participating in large projects in which such knowledge will be in demand. The impact of ongoing scientific and technical projects on the development of society and the corresponding issues of technical policy are addressed in Sect. 5.2.5.

5.2 Sociotechnical Systems

5.2.1 General Description

The term *sociotechnical system* is an extension of the concept of human–machine system to a part of society that interacts with machines at all stages of their life cycle—from design to decommissioning—or has another influence on them. The Russian national standard (GOST R 57700.3-2017) defines a sociotechnical system as a modern paradigm of "considering any production, organizational, administrative system consisting of continuous interaction of two subsystems": techno-economic (machinery, equipment, know-how, etc.) and social (humans, forms of their organization, management methods, etc.).

Sociotechnical systems are formed at the corporate, industry branch, state and transnational levels and exist during the life of one or more human generations. They function in a large-scale space and are characterized by a significant amount of resources involved and an intense impact on the life of society. Sociotechnical systems are large enterprises, organizations, energy facilities and transport infrastructure, industry branches, information and communication systems.

Such systems can be nested into each other, forming a hierarchy. An example is nuclear power. The lowest-level sociotechnical system can be considered a NPP power unit, which includes a nuclear reactor, equipment and piping components, buildings and structures, safety systems, electrical and electronic devices and communications, control system, auxiliary systems, personnel, etc. This power unit is part of a larger system—a multi-unit NPP, which is managed by the plant directorate, occupies a significant territory with restricted access and relevant infrastructure, surrounded by a sanitary protection zone and other special status zones. A nuclear power plant can be a city-forming organization: some cities in Russia were built near nuclear power plants for living the personnel with families: *Sosnovyi Bor* (Leningradskaya NPP), *Kurchatov* (Kurskaya NPP), etc. All Russian nuclear power plants are branches of JSC Concern Rosenergoatom, which, in turn, is owned by JSC Atomenergoprom (Nuclear Power Industry Complex). The latter provides a full cycle of production in the field of nuclear energy—from uranium mining to the construction of nuclear power plants and electricity generation. It unites many organizations operating in all segments of the nuclear fuel cycle and is subordinate to the State Corporation Rosatom. Control over Russian nuclear energy is also carried out at the international level: in accordance with the Treaty on the Non-Proliferation of Nuclear Weapons signed by the USSR, there is legally binding safeguards agreement with the IAEA, "in line with the IAEA's Statute" (IAEA 1998).

The properties of various sociotechnical systems, their structures and functioning are presented in the publications Mate and Silva (2005), Daft (2010), Baxter and Sommerville (2011), Ruault et al. (2012), Rettig (2017), Roth and Rebentisch (2018). The commonality of such systems is the combination of several types of interactions in them, which are considered in different scientific disciplines: physical and mechanical processes (engineering sciences, technology), information exchange and processing (information technology), interpersonal communication and interaction (psychology), social behavior (sociology). Important characteristics of any sociotechnical system are the following:

- emergence of properties—the system as a whole acquires properties only when its components are connected;
- insufficient determinism—the reaction to the same external influence is not always the same;
- subjective aspects of behavior—they manifest themselves in accordance with the views of participants in the life cycle.

Sociotechnical systems have distinctive features as compared with conventional human–machine systems: an open nature, the ability to develop and self-organize. These features are obviously due to the larger scale and wider variety of manifestations of human factors. The social subsystem can have a decisive influence on design activities. This influence may be determined by

- changes in social values over time;
- an increase of information noise;
- developments in the field of artificial intelligence;
- technical policy.

We will consider these specific manifestations of human factors sequentially.

5.2.2 Changes in Social Values

Social values are an integrated representation of people's needs and interests, ways to counter existential threats, etc. Social values include, for example, freedom of movement, safety, availability of energy, communications, economic and military power of the country. The sociotechnical systems that provide them become social values themselves. Society spends material, monetary and human resources to maintain the existence of such systems. In the creation of sociotechnical systems, the role of personalities with the property of *passionarity*[6] is great. For Soviet cosmonautics and nuclear power, such personalities were S. P. Korolyov, I. V. Kurchatov[7] and M. V. Keldysh.[8]

In the USA, cosmonautics (astronautics) has acquired social value as a technical and political response of the nation to Yuri Gagarin's space flight. In May 1961, President John F. Kennedy announced the *Apollo* program in Congress and in September 1962 presented it to the American society (Kennedy 1962). The ultimate goal was to land human on the Moon no later than 1970. This program has been conceived during the administration of Dwight D. Eisenhower, but due to his negative attitude to manned space exploration, it was being insufficiently funded. After the start of the *Apollo* program, the work was deployed throughout the country with planned government funding of $9 billion over the first 5 years. The goal was achieved on July 20, 1969: Neil Armstrong and Buzz Aldrin set foot on the surface of the Moon. In total, six successful missions were conducted under the program, the last of which took place in 1972.

Social values can change over time. For example, sociotechnical systems in the process of work harm the environment. The public's feeling of harm may increase under the influence of the mass media. In Western Europe, such feeling has developed with regard to nuclear energy and subsequently conventional thermal energy, which is generated by burning coal and hydrocarbons. The first is the object of accusations because of the risk of a radiation accident, and the second is considered responsible for global warming. As a result, the social value of energy obtained from relevant sources has decreased in comparison with the value of environmental safety.

Under pressure from environmentalists, whose interests are represented by the Green parties and expressed by many mass media, all nuclear power plants have been shut down in Italy; the governments of Belgium, Germany, Spain, Switzerland and Sweden have announced the abandonment of the use of nuclear energy. The European Union has set a binding goal of achieving climate neutrality by completely eliminating fossil fuels by 2050 (EC 2022).

---

[6] The term *passionarity* originates from the French word "passionner", which means "to captivate, excite, ignite passion". It has been introduced into Soviet historical science by Gumilyov, Lev Nikolaevich (1912–1992), the creator of the passionate theory of ethnogenesis. Passionarity implies an excess of some "biochemical energy" in a human being, which generates an irresistible inner desire for activities aimed at changing the environment and the status quo. Such activity seems to the "passionary" personality more valuable than even his own life and therefore often leads to self-sacrifice for the sake of high goals.

[7] Kurchatov, Igor Vasilievich (1903–1960) is a Soviet physicist, the chief scientific leader when solving the atomic problem in the USSR, one of the founders of the use of nuclear energy for peaceful purposes. He was the founder and first director of the Institute of Atomic Energy, now the Russian National Research Center *Kurchatov Institute*, Moscow.

[8] Keldysh, Mstislav Vsevolodovich (1911–1978) is a Soviet scientist in the field of applied mathematics and mechanics, a major organizer of Soviet science, one of the ideologists of the Soviet space program. He was the president of the USSR Academy of Sciences and led the work on the creation of Soviet computers for calculations on atomic and rocket-space programs.

Technology-induced accidents and catastrophes have a significant negative impact on the change in the public's eyes of the value of certain technologies and sociotechnical systems based on them. The abandonment of nuclear power was largely caused by the accident at the *Fukushima Daiichi* NPP in 2011—already the third in a series of similar ones. The catastrophes of the *Challenger* and *Columbia* not only led to the closure of the *Space Shuttle* program, but also caused a decrease in the interest of American society in astronautics and as a result, if not stagnation, then a significant slowdown in the development of the US space industry. The situation changed only years later—largely under the influence of the energetic activities of Elon Musk, Richard Branson and Jeff Bezos created new rockets, orbital and suborbital systems to reach space.

But such changes have not yet occurred in Western Europe. Currently, European astronauts do not have their own independent means of reaching, for example, the International Space Station. In February 2022, two of them, Thomas Pesquet and Matthias Maurer, who have flew to the ISS on Russian and American spaceships, addressed European leaders with a manifesto. This document states that the continent needs its own crewed space vehicle. If Europe misses the chance "to challenge the status quo, we will have to continue procuring human space transportation from other actors, with no guaranties that our needs and values will be a priority… We will be paying customers in a position of weakness, repeating the mistakes of the past in other strategic domains, which left us dependent on external players for our energy requirements or information technology development" (ASE 2022).

In views on the benefits and harms of certain sociotechnical systems, the interests of various groups of society may collide: some people are employees, participating in their life cycle, the owners receive income due to their social position, while others may be directly affected or belong to a competing group and try to benefit from the defeat of rivals.

The social value of high-tech industries may decline along with the deterioration of living conditions (when survival becomes the main motive for people), the decline in the level of science and culture, the loss of previous social landmarks, the degradation of state institutions responsible for technical policy and the allocation of resources for its implementation. These processes are aggravated by the insufficient development of civil society, which, being capable of self-organization, could influence politics. This is exactly how things were in the USSR in the 1990s. Not too favorable conditions for scientific and technological development in Russia existed also in the following decades. In the responses of readers to the author's article concerning current, not very good state of Russian aviation and cosmonautics (Spirochkin 2020), the main reason for this situation was noted—insufficient funding. While agreeing with the statement about the role of financing (the consideration of which was not included in the topic of this publication), it is impossible not to notice that its volume reflects, albeit indirectly, the social value of certain industries in the views of government and business. The relevant issues relate to the field of technical policy, and they are discussed in Sect. 5.2.5.

It is vitally important for specialists not to become "hostages" of changing social values, that is, not to lose their jobs when their knowledge and experience cease to be useful in the eyes of society and government representatives. Professionals should "defend" scientifically grounded views at different levels of the social subsystem with which they are connected, educate young people and influence technical policy through public organizations.

However, no one has a monopoly on the truth. The negative impact of sociotechnical systems on the environment and humans can grow slowly and be cumulative; in this case, it becomes apparent only after a long time. An example is the plastics industry. At first, plastics were considered a universal replacement for many materials, including structural ones, and their production developed rapidly. Due to their resistance to decomposition in natural conditions, they gradually littered the entire globe and turned from a social value into a tangible threat. The creators of technical systems (each of them can form a techno-economic subsystem in a sociotechnical system and thus influence the environment) should take into account such a risk, improve technologies and adapt products to natural processes.

The unpredictability of technological development, its impact on society and changes in social values can deprive many industries of prospects for existence in the near or distant future. In these conditions, it becomes important to diversify activities. Characteristic of many large companies, including multinational ones, is the simultaneous development of several technological directions. Such a strategy has been adopted in the Russian State Corporation Rosatom: the areas of supercomputer technologies, artificial intelligence and wind energy have appeared in its activities.[9]

5.2.3 Increase of Information Noise

We will consider as information noise the components present in the information that hinder its perception and understanding, distort the essence of the subject in question and (or) are do not lead to new knowledge.[10] The range of these components is very wide: from grammatical and syntactic errors to misinformation. Information noise is also created when the same information is repeatedly reproduced in different sources—due to such reproduction, new knowledge cannot arise.

For example, on one of the days when this book was being written, the Google search service gave out more than 12 million links to the author's query "Elon Musk". Each source from such a multitude, if it contains, then only a very small fraction of the original, new and useful information, which is a "signal", all the rest else is just "noise". It becomes a big challenge to select from this noise information about the true aerospace achievements of the founder, co-owner and head of SpaceX, as well as about the problems that stand in the way of his stated goal: Mars. From the standpoint of the aerospace systems designer, the most important achievement is the controlled return to Earth of the first stage of the *Falcon* 9 launch vehicle, which ensures its reuse. The booster B1060 which lifted off June 17, 2022, and carried 53

[9] According to the information from https://rosatom.ru.

[10] In the sense, discussed in Sect. 5.1.3.

Starlink satellites into space became the first *Falcon* first stage rocket to launch and land 13 times (Ralph 2022). In Russian cosmonautics, such a technique has not yet been implemented, although the relevant research and development has been carried out during the design of the *Energia-Buran* system.[11]

A serious problem of the planned Mars mission is related to the radiation that the crewmembers will be exposed to. Its main component is galactic cosmic rays, while high-energy solar particles make a smaller contribution. For 2.5 years, namely for such a period the mission is planned, the radiation dose is estimated to be at least 1 Sv with a possible increase with raising solar activity (Mewaldt et al. 2005), (Hassler et al. 2013). This value exceeds the allowable limits set by NASA, based on the 3% risk of death caused by radiation (Cucinotta et al. 2013). An astronaut in the ISS, which is largely protected from radiation by the Earth's magnetosphere, experiences a much lower radiation load. In Russian nuclear power industry, the allowable limits are even lower: the radiation dose for NPP personnel should not exceed: 0.020 Sv per year,[12] for the population in the surrounding area—0.001 Sv per year[13] (SanPiN 2.6.1.2523-09).

Over time, information noise in sociotechnical systems increases. This is facilitated by the mass media, which benefit from people's need for information or from the spread of disinformation (if it is beneficial to someone), as well as the development of information exchange technologies (from television to social networks). The latter is accompanied as a rule by a multiplication of the volume and a drop in the quality of transmitted information. Gradually, in the information exchange, the relative level of the useful "signal" decreases in comparison with the increase in "noise" generated by various kinds of interpretations, additions, repetitions and comments.

The increase in information noise absorbs material, energy and human resources, increases the risk of non-acquisition or even loss of knowledge and thereby has a negative impact on safety. Various manifestations of information noise and ways to defend against it are considered in the book Silver (2012).

A special kind of information noise arises due to the imperfection of human judgment developed when evaluating any phenomena or predicting events and processes. The risk associated with making a wrong decision based on the erroneous judgment of an expert or the opinion of a group of experts can be very high. The stochastic nature of this kind of noise, the factors affecting its level and methods of their compensation are discussed in the book Kahneman et al. (2021).

Minimization of noise components is necessary for designers when determining the design base or carrying out ergonomic design of operator workplaces, during numerical modeling and simulation (see Sects. 4.1.3 and 4.3.2) and preparation of operational documentation. It is important to manage information noise when it can affect changes in social values (Sect. 5.2.2) or lead to other costly errors (see

[11] The author was engaged in calculations to ensure the landing of the re-entry boosters of the *Energia* rocket.

[12] The average value during any consecutive 5 years, but not more than 0.05 Sv in one year.

[13] The average value during any consecutive 5 years, but not more than 0.005 Sv in one year.

above). Information noise should also be taken into account in connection with the development of artificial intelligence.

5.2.4 Development of Artificial Intelligence

The term *artificial intelligence* (AI) means:

(1) the ability of computer systems to perform intellectual functions inherent in humans;
(2) the field of science and technology related to the creation of such systems.

Intellectual functions intended to replace the work of the human brain are implemented by sophisticated software programs. The question of whether computers equipped with appropriate programs are capable of mental activity belongs to Alan M. Turing, who formulated it in 1950 (Turing 1950). However, to this day, more than 70 years later, there is no exact criterion, the satisfaction of which would indicate the presence of intelligence in a computer system. The state of research and development in the field of AI at the beginning of the twenty-first century is presented, for example, in the book Russell and Norvig (2021). This field includes expert systems, knowledge bases and logical inference systems, neural networks, pattern recognition systems, etc. The distinctive features of "intelligent" software tools (among other software) are, apparently, the possibilities of machine learning and decision-making in conditions of uncertainty.

The current level of AI implementation in technical systems is determined by the use of some of its elements in robotics, unmanned vehicles, security systems and a number of other areas. An idea of the ongoing work and its financing can be obtained, for example, from the report MGI (2017). The International Organization for Standardization prepares standards for various applications of artificial intelligence. For example, in 2019–2021, the following standards were issued[14]:

- ISO/IEC 20546:2019 Information technology—Big data—Overview and vocabulary;
- ISO/IEC TR 20547-1:2020 Information technology—Big data reference architecture—Part 1: Framework and application process;
- ISO/IEC TR 20547-3:2020 Information technology—Big data reference architecture—Part 3: reference architecture;
- ISO/IEC TR 24027:2021 Information technology—Artificial intelligence (AI)—Bias in AI systems and AI aided decision making;
- ISO/IEC TR 24028:2020 Information technology—Artificial intelligence—Overview of trustworthiness in artificial intelligence;
- ISO/IEC CD TR 24029-1 Artificial Intelligence (AI)—Assessment of the robustness of neural networks—Part 1: Overview;

[14] According to the information from https://www.iso.org/standard/77608.html.

- ISO/IEC TR 24030:2021 Information technology—Artificial intelligence (AI)—Use cases;
- ISO/IEC TR 24372:2021 Information technology—Artificial intelligence (AI)—Overview of computational approaches for AI systems.

The following documents are under development:

- ISO/IEC DTS 4213.2 Information technology—Artificial Intelligence—Assessment of machine learning classification performance;
- ISO/IEC AWI 5259-1 Artificial Intelligence—Data quality for analytics and machine learning (ML)—Part 1: Overview, terminology, and examples, ISO/IEC AWI 5259-2 …—Part 2: Data quality measures, ISO/IEC AWI 5259-3 …—Part 3: Data quality management requirements and guidelines, ISO/IEC AWI 5259-4 …—Part 4: Data quality process framework, ISO/IEC AWI 5259-5 …—Part 5: Data quality governance;
- ISO/IEC CD 5338 Information technology—Artificial intelligence—AI system life cycle processes;
- ISO/IEC AWI 5339 Information technology—Artificial intelligence—Guidelines for AI applications;
- ISO/IEC AWI 5392 Information technology—Artificial intelligence—Reference architecture of knowledge engineering;
- ISO/IEC AWI TR 5469 Artificial intelligence—Functional safety and AI systems;
- ISO/IEC AWI TS 5471 Artificial intelligence—Quality evaluation guidelines for AI systems;
- ISO/IEC AWI 12792 Information technology—Artificial intelligence—Transparency taxonomy of AI systems;
- ISO/IEC FDIS 22989 Information technology—Artificial intelligence—Artificial intelligence concepts and terminology;
- ISO/IEC DIS 23894 Information Technology—Artificial Intelligence—Risk management;
- ISO/IEC DTR 24368 Information Technology—Artificial Intelligence—Overview of ethical and societal concerns, etc.

The last of the listed standards seems particularly noteworthy, since, judging by its name, it concerns an area that is considered so far exclusively inherent in human.

In Russia, the national standard (GOST R 58777-2019) has been enacted; it concerns the use of AI elements in the recognition of X-ray images by technical means of inspection at airports. As for other practical applications, some of them relate to Russian nuclear power industry (Dorohova 2020):

- visual monitoring of the use of protective equipment by personal and detection of violations in real time (the experiment is carried out at the Kola NPP);
- content search in the documentary archives of JSC Concern Rosenergoatom for descriptions of equipment failures at nuclear power plants in order to identify risk factors and prevent unplanned downtime;
- simultaneous translation for training personnel of nuclear power plants under construction abroad.

Ideas and achievements in the field of AI are accompanied by significant information noise (see Sect. 5.2.3), which should be borne in mind when analyzing publications. The existing AI elements function mainly on the basis of statistical data processing. As a result of such processing, correlations can be detected, but the computer identification of cause-and-effect relationships and the formation of contextual understanding (characteristic of the human mind) are not yet possible. Developers of AI do not succeed to implement the emergence of knowledge in the computer.

The reason is, apparently, the fundamental limitation of the capabilities of artificial intelligence associated with the Gödel's incompleteness theorems. The consequence of these theorems is the incompleteness of any formal system built on axioms. This may mean that in every axiomatic system, including mathematics or logic, there is a part, rather implicit, that is not derived from axioms, but originates from physical experience. Perhaps, it is this part that forms the context necessary for the production of knowledge. And physical experience is not available to computers—at least not yet. This limitation of artificial intelligence is noted in the article related to one of its sections—neural networks (Colbrook et al. 2022).

For designers, the following is important: the principal possibility of quickly and reliably solving certain tasks with the help of artificial intelligence, determining the requirements for reproducing some functions of the human mind in computers, ensuring control over the operation and support of software and hardware used during the life cycle of machines equipped with them. One of the tasks that is supposed to be "entrusted" to AI is the recognition of critical situations during the operation of a technical system, for example, an airplane (at a speed exceeding human capabilities) and their control (see Sect. 4.2.2). However, since there is no understanding yet of how the mechanisms of flexible human logic work, this logic cannot be reproduced completely in computer software. Therefore, at the current stage of development, the use of AI to control situations with the possibility of catastrophic consequences is limited.

In connection with the implementation of AI in sociotechnical systems, certain threats arise, which include the following:

- the possibility of introducing hard-to-detect errors into the developed intellectual software tools (due to human factors inherent in the developers of such tools);
- loss of competence and understanding of applied technologies by operating personnel if they are excluded from the life cycle processes of techno-economic subsystems;
- degradation of human thinking abilities at the mass level (Kurpatov 2018; Kurpatov 2020), a kind of "digital idiocy".

In Sect. 5.1.3, it was already mentioned that experts noticed a decrease in humans' mental abilities as a result of excessive use of information technology. The corresponding influence is also produced by information noise (see Sect. 5.2.3), but the key contribution to this negative process may be made by elements of artificial intelligence, which consistently replace the work of the human mind to solve intellectual problems. An important observation contained in the book of Jared Diamond, a researcher who worked in New Guinea, (Diamond 1997), should be mentioned here. It concerned the greater development of mental abilities among representatives of tribes that are considered to be wild and undeveloped (since they did not use the achievements of technological civilization, including mass media, until recently) than among modern Western Europeans and Americans (immersed from birth in technology, including information one). The cognitive abilities of the supposedly "savages", which attracted the attention of Jared Diamond, were being developed in an unfavorable environment over thousands of years and have had existential significance.

The author, of course, in no case calls for stopping the development of technology, and especially artificial intelligence. It would be stupid and simply impossible. But understanding the prospects for the development and potential threats from artificial intelligence is necessary for the formation of appropriate technical policy at the level of organizations engaged in relevant R&D, industries and government agencies.

5.2.5 Technical Policy

By technical policy, we will understand the management of sociotechnical systems to achieve goals related to the life cycle of the machines included in these systems. Such management covers:

- setting goals;
- ensuring their achievement with resources;
- evaluating obstacles or disturbances that may arise outside and inside the system and threaten the achievement of goals;
- determining the control actions necessary to eliminate obstacles or compensate for disturbances, and ensuring such actions;
- analyzing the results of this management and the situation (changing over time) as a whole.

For the sustainable existence and development of a sociotechnical system, it is important that the set goals, allocated resources and management methods correspond to the inherent properties of the system that emerged at its creation (see Sect. 5.2.1) and change during the life cycle. These properties are determined, among other things, by the participation of people who make up the social subsystem, with all possible manifestations of human factors. Due to insufficient determinism in the characteristics of sociotechnical systems (Sect. 5.2.1), as well as in the external environment, the applicability of mathematical methods of automatic control theory is limited. Therefore, neither disturbances, nor control actions, nor the required resources can be forecasted with the necessary accuracy.

The uncertainty inherent in sociotechnical systems, along with their fundamental complexity, devalues long-term development programs presented by managers at various levels as technical policies. A rational approach consists only in setting goals and focusing the activities of the participants in the life cycle of the system on ensuring such properties that contribute to sustainability and development, namely:

- negative feedback;
- connectivity of components horizontally (not just hierarchically);
- autonomy of parts;
- duplication of key functions;
- exclusion of cliff-edge effects (that hinder management and can lead to irreversible consequences);
- self-healing ability, etc.

Together, these properties should provide an integral quality of system resilience that corresponds to modern views on safety. A more complete set of characteristics that ensure resilience of sociotechnical systems as well as a bibliography on this topic can be found in the publication Ruault et al. (2012).

Resilience is an extension of the concept of sustainability and, like this initial concept, can obviously be implemented only within certain limits. Due to the uncertainty and complexity of sociotechnical systems, it is difficult to determine these limits in advance. They may change over time due to aging and obsolescence of machinery components (see Sects. 5.1.1 and 5.1.2), as well as changes in the social subsystem. These changes reflect the manifestations of human factors as the loss of design knowledge (Sect. 5.1.3), the transformation of social values (Sect. 5.2.2) and the increase of information noise (Sect. 5.2.3). The introduction of artificial intelligence also affects changes in the social subsystem (Sect. 5.2.4). With such a multitude of factors that cannot be accurately determined, reliable forecasting of the development of a sociotechnical system is impossible.

Within the framework of technical policy, dynamic management is necessary, including monitoring of the changing situation and ongoing evaluation of the results of control actions. The "relationship" of specialists with technical policy is similar to that with social values (see Sect. 5.2.2): the possibility and content of their professional activity largely depend on external factors. The latter include decisions made by superior managers based on their position in the hierarchy, available resources, competence, understanding of responsibility, views formed by the "inner circle" and mass media, personal qualities, interests, etc. Specialists are able to influence decisions in the organization, less often in the industry branch, but only in exceptional cases—at the state level. Due to the scale and inertia of sociotechnical systems, the results of technical policy are fully manifested only after a long time.

An example of a state technical policy, the results of which cannot yet be considered negative, but which are becoming more and more doubtful, is the refusal of a number of countries from nuclear energy. In particular, the German government is being severely criticized because of the decision to stop the operation of all nuclear power plants in the country in 2022 (Bethge 2019; Bittner 2020). This decision was based on mass anti-atomic sentiments, which leave no chance for the transition, for example, to more modern and safe closed fuel cycle technologies using fast neutron reactors, small modular nuclear power plants, etc. Such technologies are considered in the part of the world where they are being researched, developed and applied (including Russia) as the best alternative to the conventional thermal energy generation and an ideal means of combating climate change. It is quite possible that due to the complete abandonment of nuclear energy, Germany is missing out on technological development opportunities to a greater extent than it prevents dangers.

An example of a technical policy, the negative consequences of which have already become obvious to almost everyone, is the policy pursued in Russian civil aviation and cosmonautics. In the USSR, there was a powerful aviation industry that produced, along with military airplanes and helicopters, various types of civilian aircraft. The Soviet Union has launched the first artificial satellite of the Earth, *Sputnik* (1957), and has become the first country to send a man into space (1961). Although 7 years late (compared to American space shuttles), the Soviet space transport system *Energia-Buran* has been created and two test launches have been carried out. The *Energia* launch vehicle had more versatility of application than space shuttles since its central unit had its own propulsion system. This rocket could deliver various payloads to the Earth orbit, not only in the payload bay of the Soviet space shuttle *Buran* (see Fig. 5.3) and not only *Buran* itself—in this case, the payload mass could be more than 100 tons (for American space shuttles, it was about 30 tons). *Buran* was designed, unlike its American prototype, with the possibility of unmanned flight (in automatic mode) and safe landing in case of landing gear failure (see Sect. 2.4.3).

As a result of the collapse of the USSR and the socioeconomic reforms of the 1990s, the degradation of the Soviet scientific and technical potential occurred. Currently, civil air vehicles being operated in Russia are mainly of foreign production. The share of domestic airplanes is no more than 25%, among them modern models—no more than 10% (Nikol'skaya et al. 2015), (Air Charter Service 2019). From 2011 to 2016 inclusive, the State Corporation Roscosmos was the undisputed

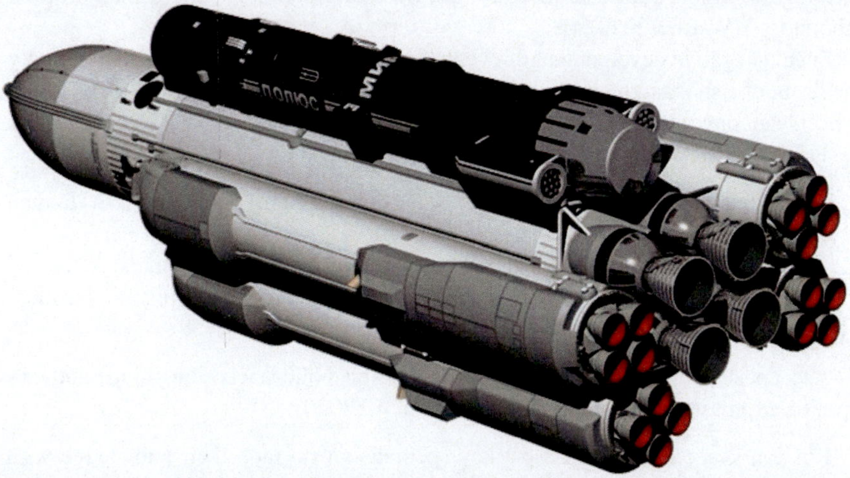

Fig. 5.3 Russian launch vehicle *Energia* with experimental payload *Polyus*. Reproduced from https://www.buran.ru/htm/36-3.htm with the permission of Dr. Vadim P. Lukashevich

Table 5.1 Number of civil space launches in 2017–2021, according to data from Sidorkova et al. (2018), The Universe (2020), Kolpaksidi (2020), Sozinov (2022)

Year	Number of civil space launches			
	Russia	PRC	USA	
			In total	By SpaceX: Falcon 9 + Falcon Heavy
2017	15 (1)*	10 (2)	23	16
2018	17 (1)	39 (1)	31	20 + 1
2019	22	34 (2)	21	11 + 2
2020	15	39 (4)	37 (3)	25
2021	24 (1)	55 (3)	45 (2)	31

*Of them unsuccessful, in whole or in part

world leader in the number of civilian space launches. But then the situation began to change, as Table 5.1 demonstrates. In 2017, the private American company SpaceX was ahead of Roscosmos in successful launches of this kind: 16 versus 15. In 2018, Russia was overtaken by the People's Republic of China (38 vs. 16). And although in 2019 and 2020 all Russian launches were accident-free, the lag in their number from competitors increased. In 2021, Russia's lag in space activities from China and the USA remained.

In connection with the missions of manned *Crew Dragon* spaceships to the ISS, the first of which took place in the summer of 2020, Russian cosmonautics faces defeat in competition with not only the state space agencies of these countries, but also private business represented by SpaceX. After many years of dependence on Russian *Soyuz* spacecraft delivering crews to the ISS, NASA became the opportunity to carry out its own regular flights and "take more control over US operations on the giant orbiting laboratory" (Wattles 2020).

Such a negative development of events for Russian cosmonautics is undoubtedly the result of a short-sighted technical policy. It can be characterized by the following aspects demonstrated over the past decades:

(1) insufficient attention of the state to this industry (lack of reasonable goals, insufficient funding for research and development, inattention to human resources);
(2) incompetence of management and inefficient use of allocated funds;
(3) lack of public–private partnership (including support for private business initiatives), etc.

As a consequence of the two first aspects of this policy, according to the author's opinion expressed in the publication Spirochkin (2020),

- Russian space industry has lost key specialists who took their knowledge with them;
- the younger designers have not acquired the necessary skills to solve complex technical problems due to the gap of continuity between generations.

The third of listed aspects is demonstrated, for example, by the sad fate of the Russian solar sail spaceship project. The design of this unusual spaceship, the prototype of which was tested in 1993 (the *Znamya-2* space experiment), was an undoubted achievement not only of late Soviet and Russian, but also of world cosmonautics. But in the following years, this project was practically buried in the bureaucratic jungle of the Russian space authority (Spirochkin and Saprykin 2021).

As for the inattention to human resources (and it is they who embody human factors), these resources are very important for American colleagues. For example, FAA and the Air Traffic Control Association (ATCA) are concerned about how to inspire the younger generation to pursue a career in the aerospace industry and pass on knowledge to them. These organizations support initiatives of dialog with young people, organize events to inform and train them with the participation of professionals (FAA 2019; ATCA 2020). The declining state of Russian civil aviation and cosmonautics is probably affected also by the absence of "passionary" personalities (the term discussed in Sect. 5.2.2) who, having material resources, would have a vision of prospects and the necessary personal qualities to overcome technical and bureaucratic obstacles.

A contrasting example of a successful technical policy is demonstrated by Russian nuclear power industry. It would seem that after the Chernobyl disaster, the change of socioeconomic formation and the general decline in the country, it was threatened with imminent collapse. The construction of new nuclear power plants has stopped; design institutes and the entire industry have found themselves in a hard situation. This situation was saved by certain manifestations of human factors at different levels of this large sociotechnical system. Perhaps, the initiative "from below" was crucial, which came from a group of SPbAEP[15] specialists who performed unscheduled and unpaid work on preparing a proposal for entering the foreign market. It was not without simple luck, but the then Ministry of Atomic Energy of Russia supported the initiative and arranged a meeting of representatives of this design institute with a potential customer from PRC. As a result, the first two power units of the *Tianwan* NPP were designed and built a few years later.

The specialists of the Moscow AEP,[16] who were designing the Novovoronezhskaya NPP-2, followed the same path. Nuclear power industry began to revive in Russia, and the spirit of the "atomic renaissance" arose. Currently, the State Corporation Rosatom ranks first in the world on the portfolio of projects implemented abroad: 36 nuclear power units in 12 countries are in different project stages.[17] Three units are being built in Russia. On May 22, 2020, the floating nuclear power plant *Akademik Lomonosov* was commissioned; it became the 11th nuclear power plant in Russia and the northernmost in the world.

---

[15] The St. Petersburg Research and Design Institute ATOMENERGOPROEKT. Now, JSC Atomproekt, one of the principal design organizations in Russian nuclear industry.

[16] Now, JSC Atomenergoproject.

[17] According to the information from https://rosatom.ru.

Conclusion

The author sought to show in this book how human factors are taken into account and managed in order to ensure the safe operation of technical systems. The level of accounting and management is characterized by developed regulatory documents, standardized design and construction solutions, organizational measures, a variety of applied information and mathematical models, a large amount of accumulated experimental data. The author hopes that the information presented will allow the reader to navigate the vast and complex area of human–machine interaction. For more detailed information, the reader can refer to the sources listed in Bibliography.

The main attention was paid to the design stage, during which the configuration, structure and functional scheme of the technical system are created, and its efficient and safe operation is ensured. The features inherent in humans not only affect the properties that designers endow the system with, but are also reflected in the design process itself. The completeness of how human factors are taken into account when designing and how design activities are managed in order to minimize the negative manifestations of these factors became the subject of analysis, which is not often found in scientific and technical publications. This situation is understandable: most of the articles and monographs addressed to engineers and researchers relate to the applied problems they solve. It is difficult to rise above the flow of daily emerging issues related to the development of structure, equipment, hardware and software, design calculations, tests, processing of their results and other current affairs—to see the fundamental problems caused by human participation in the life cycle of the created technical systems. Modern managers who have to control the design processes, solve such problems and simultaneously eliminate "traps" at the organizational level are not always able to bridge the gap between the socioeconomic content of management (which has long been formed into a separate profession) and engineering disciplines that are the foundation of design. Such gap as well as any other disruption of continuity leads to poorly predicted consequences. When analyzing human–machine interaction, the author tried to minimize this gap, relying on his knowledge and experience of working at different levels of the hierarchy in

Y. Spirochkin, *Human Factors and Design*,
https://doi.org/10.1007/978-981-19-8832-5

two high-tech industries: aerospace and nuclear. But the differences in the activities of representatives of engineering and management, of course, are great; they are reinforced by the specifics of various fields of engineering. Therefore, there is a risk that part of what is presented in the book will be difficult for narrow specialists to understand. Many technical problems are interconnected and interdisciplinary in nature. Mastering new adjacent knowledge can help in solving them. Familiarization with approaches peculiar to other professions and achievements of other industries can be useful and can give a synergistic effect.

The analysis carried out in the book showed that in the complex of safety measures, generally reliable, there are lacunas due to insufficient consideration of all hazards associated with human factors. Serious challenges arise not only because of human errors and the unpredictability of human behavior, but also because of the uncertainty and complexity inherent in design activities. The book offers possible answers to existing challenges, including:

- implementation of the concepts of Safety II (preparedness for unpredictable events) and Resilience engineering (creation of sustainable, self-recovering systems) in the design;
- a shift in the design paradigm toward increasing the importance of safety compared to economic indicators (which now dominate in many cases);
- the need to reflect in regulatory documents the variability of personnel actions (including the possibility of inaction) and various forms of behavior of the public in danger conditions;
- creation of protection that prevents the control of the system by an attacker, including that from personnel, etc.

An additional challenge and, accordingly, a risk factor is the increasing complexity of the mathematical models used in the design. Numerical models are intended to effectively solve engineering problems, but at the same time, they create a new problem. It seems to the author that complexity of the models is starting to get out of control under the influence of persistent (if not aggressive) promotion of commercial software. This jeopardizes the development and application of analytical skills, independent of information technology, which are necessary for designers. Instead of uncontrolled complication, regulation is required in order to rationally combine analytical and numerical analysis tools and thus retain invariant design knowledge.

The design problems associated with unpredictable events are only partially covered in the book. New phenomena arise in life, the impact of which on further technical development is difficult to predict. While writing the Russian-language manuscript of this book, the COVID-19 coronavirus pandemic began to spread in the world. The discrepancy between the current principles of designing the transport systems and infrastructure facilities and the changed conditions of safe existence has become obvious. Our trains, planes and buildings were not designed to provide a distance between people from 1.5 to 2 m. Their design did not take into account the possibility of large-scale spread of infection in a confined space, through railings, door handles and air-conditioning systems. Even now, more than two years after the start of the pandemic, when it has waned and lessons can be learned, it is not clear how the new realities will be taken into account in the future projects. However, such accounting is necessary. Some researchers have long considered viral pandemics to be the main danger to humanity (Bostrom and Cirkovic 2008).

The development of artificial intelligence is also difficult to predict. AI exists in the medium formed of computers and communications; more and more functions related to the design, manufacturing and operation of technical systems are gradually transferred to electronic "brains" and "hands". So far, computers are not able to determine the requirements for the designed system and draw conclusions related to the field of knowledge. However, the creators of AI are working intensively in this direction. The results of this work will contribute in increasing the safety and comfort of people by excluding human factors with their possible negative manifestations from the processes of the life cycle of systems. But is there any guarantee that removing a human from solving problems will not lead to atrophy of thinking abilities? Where is the "red line" in minimizing the influence of human factors on further technical development that should not be crossed?

The book does not address the problems of designing biotechnical devices that interact with humans on a micro- or more subtle level. These include, for example, sophisticated medical instruments and nanorobots designed to work inside the human body. The principles of their creation, taking into account the human–machine interaction on such a small spatial scale, obviously should differ from those described above.

At the same time, the methodological basis, including theoretical, computational and experimental methods of substantiating design and construction decisions, is likely to remain the same. Some ideas about such problems and ways to solve them, for example, in the field of eye microsurgery, can be obtained in the author's publication (Spirochkin 2013). It seems that approaches to human factors in biotechnical design on this level are still waiting for their researchers and developers.

In the end, the author would like to thank the people with whom he discussed the contents of this book, who made comments and recommendations, provided information assistance.

The author owes the idea of this book to his acquaintance with Valery G. Shelkovnikov, a veteran of civil aviation, now President of the International Consultancy and Analysis Agency *Aviation Safety*. The mutual desire to find as many points of contact as possible between our areas of interest led to their discovery in the field of human factors. Many ICAO documents and other information resources of the world aviation community, including the Flight Safety Foundation and ATCA, of which Mr. Shelkovnikov is a respected member, have become available to the author.

Specialists in nuclear energy, who previously worked in this industry, and those who remain "in service" to this day, shared with the author their thoughts about its past and current state. Unfortunately, not all of them wanted to be mentioned in the book.

When writing the Russian-language manuscript, the author discussed professional issues with Dimitar Popov, a project engineer at AREVA, involved in the construction of the *Olkiluoto* 3 nuclear power plant in Finland. The prepared text was carefully read by him, and throughout the work on the book, the author felt his friendly support.

Dr. Manfred Siakkou drew the author's attention to publications in the German media about the attitude of residents of this country and some other EU countries to nuclear energy. Conversations with him helped the author to formulate a more reasonable presentation of both these and other sociotechnical issues that are controversial and cause discussions in society.

Comments on the manuscript were made by colleagues working in Russian space industry: Igor V. Pletnyov and Sergey N. Atroshenkov (JSC S. P. Korolyov Rocket and Space Corporation *Energia*), as well as by Dr. Oleg A. Saprykin, General Director of the JSC Consortium *Space Regatta.*

During the long and sometimes tedious preparation of the English-language edition of the book, the author relied on the advice of his old friend Dr. Gerhard Krause from Potsdam (Germany), his empathy and valuable organizational help.

This preparation would hardly have been successfully completed without the consulting assistance, professionalism and perseverance of Elena G. Popova.

The author is grateful to all these people for their attention to his work and contribution to the publication of the book.

Bibliography

Abramowicz W (2001) Macro element method in crashworthiness of vehicles. In: Ambrósio JAC (ed) Crashworthiness. Energy management and occupant protection, Part II. Springer, Wien, New York, pp 83–138

Abramowitz A, Smith TG, Vu T (2000) Vertical drop test of a narrow-body transport fuselage section with a conformable auxiliary fuel tank onboard. In: Final report. DOT/FAA/AR-00/56. FAA William J. Hughes Technical Center, Atlantic City International Airport, NJ. U.S. Department of Transportation, FAA, Washington, DC

Abramowitz A, Vu T (2008) Vertical impact response characteristics of four commuter/regional airplanes. In: Final report. DOT/FAA/AR-08/20. FAA William J. Hughes Technical Center, Atlantic City International Airport, NJ. U.S. Department of Transportation, FAA, Washington, DC

AC 29-2C (2018) Advisory circular (with changes 1-8). Federal Aviation Administration, 2 July 2018. https://www.faa.gov/documentLibrary/media/Advisory_Circular/AC_29-2C_with_changes_1-8.pdf. Accessed 24 March 2022

ACAS Guide (2017) Airborne collision avoidance. European Organisation for the Safety of Air Navigation (EUROCONTROL), Brussels. https://www.eurocontrol.int/sites/default/files/2019-03/safety-acas-2-guide.pdf. Accessed 24 March 2022

Airbus (2020) Airbus UTM. Deploying unmanned traffic management solutions. Airbus S.A.S., https://www.airbus.com/en/innovation/autonomous-connected/unmanned-traffic-management/airbus-utm. Accessed 24 March 2022

Akopov AS, Beklaryan LA (2012) Simulation of human crowd behavior in extreme situations. Int J Pure Appl Math 79(1):121–138

Ambrósio JAC, Pereira MS (1997) Multibody dynamic tools for crashworthiness and impact. In: Ambrósio JAC, Pereira MS, da Silva FP (eds) Crashworthiness of transportation systems: structural impact and occupant protection (NATO ASI Series. Series E: Applied Sciences, Volume 332), Part IV. Springer, Science & Business Media, Dordrecht, Netherlands, pp 475–520

Ambrósio JAC, Silva MPT (2001) Vehicle and occupant integrated simulation. In: Ambrósio JAC (ed) Crashworthiness. Energy management and occupant protection, Part IV, Chapter 18. Springer, Wien, New York, pp 263–280

Ambrósio JAC, Dias JP, Pereira MS (2001) Advanced design of structural components for crashworthiness. In: Ambrósio JAC (ed) Crashworthiness. Energy management and occupant protection, Part IV, Chapter 19. Springer, Wien, New York, pp 281–302

Anderson JD Jr (2010) Aircraft performance and design. Tata McGraw Hill, Delhi

Y. Spirochkin, *Human Factors and Design*,
https://doi.org/10.1007/978-981-19-8832-5

Archibald RD (2008) Managing high-technology programs and projects, 3rd edn. John Wiley & Sons, New York

Armscom (2017) Art of defense information. News & Press releases. Undersea rescue command installs submarine rescue diving, recompression system. 01 May 2017. http://armscom.net/news/undersea_rescue_command_installs_submarine_rescue_diving_recompression_system. Accessed 23 March 2020

ASE (2022) Association of space explorers—Europe. European astronauts' manifesto on the occasion of the European Space Summit. https://www.space-explorers.org/resources/Documents/ASE_Manifesto.pdf. Accessed 19 Feb 2022

ASME (2019) American Society of Mechanical Engineers. ASME Boiler & Pressure Vessel Code. 2019 Edition. An International Code. ASME, New York

ASN (2002) Aviation Safety Network—ASN Accident Database. https://aviation-safety.net/database/record.php?id=20020701-0. Accessed 3 June 2022

ASN (2015) Aviation Safety Network—ASN News. List of aircraft accidents and incidents intentionally caused by pilots. http://news.aviation-safety.net/2015/03/26/list-of-aircraft-accidents-and-incidents-deliberately-caused-by-pilots. Accessed 17 March 2020

ASN (2019) Aviation Safety Network—ASN Accident Database. https://aviation-safety.net/database/record.php?id=20190815-0. Accessed 1 Dec 2019

Asselin M (2012) An introduction to aircraft performance (AIAA Education Series). American Institute of Aeronautics and Astronautics, Reston, VA

ATCA (2020) Air Traffic Control Association. Young aviation professionals. https://www.atca.org/youngprofessionals. Accessed 20 July 2020

Austin R (2010) Unmanned aircraft systems. UAVS design, development and deployment. John Wiley & Sons, Chichester, UK

Bathe KJ, Wilson EL (1976) Numerical methods in finite element analysis. Prentice-Hall, Englewood Cliffs, NJ

Baxter G, Sommerville I (2011) Socio-technical systems: from design methods to systems engineering. Interact Comput 23(1):4–17

Baykasoğlu C, Mugan A, Sunbuloglu E, Bozdag SE, Aruk F, Toprak T (2013) Rollover crashworthiness analysis of a railroad passenger car. Int J Crashworthiness 18(5):492–501

Beheshti HK, Lankarani H (2010) An investigation in crashworthiness evaluation of aircraft seat cushions at extreme ranges of temperature. J Mech Sci Technol 5(24):1105–1110

Bethge P (2019) Strahlend grün. Der Spiegel, 14. Dezember 2019, Nr. 51:112–118

Bittner J (2020) The tragedy of Germany's energy experiment. The New York Times, 8 Jan 2020

Bolukbasi A, Crocco J, Clarke C, Fasanella E, Jackson K, Keary P, Labun L, Mapes P, McEntire J, Pellettiere J, Pilati B, Rumph F, Schuck J, Schultz M, Smith M, Vasquez D (2011) Full spectrum crashworthiness criteria for rotorcraft. Final report. RDECOM TR 12-D-12. U.S. Army Research, Development & Engineering Command. Aviation Applied Technology Directorate, Fort Eustis, VA

Bostrom N, Cirkovic MM (eds) (2008) Global catastrophic risks. Oxford University Press, Oxford

Byar A, Awerbuch J, Lau A, Tan T (2001) Finite element simulation of a vertical drop test of a Boeing 737 fuselage section. In: Proceedings of The Third Triennial International Fire & Cabin Safety Research Conference. Atlantic City, NJ, 22–25 Oct 2001. https://www.fire.tc.faa.gov/2001Conference/html/crash/simulation.htm. Accessed 28 March 2022

CAAP 71 (2016) Civil Aviation Advisory Publications. Helidecks (Off-Shore). Standards, guidance and information regarding helidecks, Issue 02. General Civil Aviation Authority, United Arab Emirates

CAIB (2003) Columbia Accident Investigation Board. NASA. Space Shuttle Columbia and her crew. Report of Columbia Accident Investigation Board, Volume I. Chapter 6. Decision making at NASA. https://www.nasa.gov/columbia/home/CAIB_Vol1.html. Accessed 29 March 2022

CFR Part 25. Code of Federal Regulations/Title 14/Chapter I/Subchapter C/Part 25—Airworthiness Standards: Transport Category Airplanes. https://www.ecfr.gov/current/title-14/chapter-I/subchapter-C/part-25. Accessed 1 March 2022

CFR Part 73. Code of Federal Regulations/Title 10/Chapter I/Part 73—Physical Protection of Plants and Materials. https://www.ecfr.gov/current/title-10/chapter-I/part-73. Accessed 1 March 2022

CFR Part 238. Code of Federal Regulations/Title 49/Subtitle B/Chapter II/Part 238—Passenger Equipment Safety Standards. https://www.ecfr.gov/current/title-49/subtitle-B/chapter-II/part-238. Accessed 1 March 2022

CFR Part 571. Code of Federal Regulations/Title 49/Subtitle B/Chapter V/Part 571—Federal Motor Vehicle Safety Standards. https://www.ecfr.gov/current/title-49/subtitle-B/chapter-V/part-571. Accessed 1 March 2022

CFR Part 572. Code of Federal Regulations/Title 49/Subtitle B/Chapter V/Part 572—Antropomorphic Test Devices. https://www.ecfr.gov/current/title-49/subtitle-B/chapter-V/part-572. Accessed 1 March 2022

CFR Part 585. Code of Federal Regulations/Title 49/Subtitle B/Chapter V/Part 585—Phase-in Reporting Requirements. https://www.ecfr.gov/current/title-49/subtitle-B/chapter-V/part-585. Accessed 1 March 2022

Chapanis A, Garner WR, Morgan CT (1949) Applied experimental psychology: human factors in engineering design. John Wiley & Sons, New York

Chapanis A (1965) Man-machine engineering. Wadsworth Publishing Company, Belmont, CA

Clifford J (2020). Total human model for safety: What is THUMS? Toyota UK Magazine, 18 June 2020. https://mag.toyota.co.uk/thums-total-human-model-for-safety. Accessed 25 March 2022

Colbrook MJ, Antun V, Hansen AC (2022) The difficulty of computing stable and accurate neural networks: on the barriers of deep learning and Smale's 18th problem. Proc Natl Acad Sci 119(12)

Crawford M (2011) Facial recognition progress report. SPIE. The International Society for Optics and Photonics, 28 Sept 2011. https://spie.org/news/facial-recognition?SSO=1. Accessed 25 Feb 2020

CS-23 (2003) Certification specifications for normal, utility, aerobatic, and commuter category aeroplanes. European Aviation Safety Agency, Brussels

CS-25 (2018) Certification specifications for large aeroplanes (Amendment 21). European Aviation Safety Agency, Brussels

Cucinotta FA, Kim M-HY, Chappell LJ (2013). Space radiation cancer risk projections and uncertainties—2012. NASA/TP-2013-217375. NASA Center for AeroSpace Information, Hanover, MD

Daft RL (2010) Organisation theory and design, 10th edn. South-Western, Cengage Learning, Mason, OH

Deac SC, Perescu A, Simoiu D, Nyaguly E, Crâştiu I, Bereteu L (2017) Modelling and simulation of cars in frontal collision. IOP Conf Series: Mater Sci Eng 294, 012090. IOP Publishing, Bristol, UK

Dédale JP (2013) Why a paradigm shift is needed. IAEA International Experts' Meeting on Human and Organizational Factors in Nuclear Safety in the Light of the Accident at the Fukushima Daiichi Nuclear Power Plant. Vienna, Austria, 21–24 May 2013. Conference ID: 45441 (I2-CN-212). https://www-pub.iaea.org/iaeameetings/IEM5/IEM5_Jean%20Paries_Dedale_France.pdf. Accessed 29 March 2022

Dekker K, Verwer JG (1984) Stability of Runge-Kutta methods for stiff nonlinear differential equations. North-Holland, Amsterdam, New York, Oxford

Diamond J (1997) Guns, germs, and steel: the fates of human societies. W. W. Norton & Company, New York

Du Bois P, Chou CC, Fileta BB, Khalil TB, King AI, Mahmood HF, Mertz HJ, Wismans J (2000) Vehicle crashworthiness and occupant protection. (Editors: Prasad P, Belwafa JE). American Iron and Steel Institute, Southfield, MI

EASA (2020) European Union Aviation Safety Agency. Opinion No 01/2020. High-level regulatory framework for the U-space. RMT.0230. https://www.easa.europa.eu/sites/default/files/dfu/Opinion%20No%2001-2020.pdf . Accessed 16 July 2020

EC (2022) European Council | Council of the European Union. European Green Deal. Fit for 55. The EU's plan for a green transition. https://www.consilium.europa.eu/en/policies/green-deal/fit-for-55-the-eu-plan-for-a-green-transition. Accessed 21 July 2022

EN 1317-1:2010. European Standard. Road restraint systems—Part 1: Terminology and general criteria for test methods. European Committee for Standardization, Brussels

EN 1317-2:2010. European Standard. Road restraint systems—Part 2: Performance classes, impact test acceptance criteria and test method for safety barriers including vehicle parapets. European Committee for Standardization, Brussels

EN 13445-1:2009. European Standard. Unfired pressure vessels—Part 1: General. European Committee for Standardization, Brussels

EN 15227:2008 + A1:2010. European Standard. Railway applications—Crashworthiness requirements for railway vehicle bodies. European Committee for Standardization, Brussels

FAA (2001) Federal Aviation Administration. Aircraft Crashworthiness Research Program. FAA William J. Hughes Technical Center, Atlantic City, NJ. http://www.tc.faa.gov/its/cmd/visitors/data/AAR-430/crashwor.pdf. Accessed 1 Jan 2020

FAA (2011) Federal Aviation Administration. Introduction to TCAS II. Version 7.1. February 28, 2011. U.S. Department of Transportation, FAA, Washington, DC. https://www.faa.gov/documentlibrary/media/advisory_circular/tcas%20ii%20v7.1%20intro%20booklet.pdf. Accessed 30 March 2022

FAA (2019) Federal Aviation Administration. Section 601. Youth in aviation student outreach report. U.S. Department of Transportation, FAA, Washington, DC

FAA (2020) Federal Aviation Administration—Unmanned Aircraft Systems—Research and Development—UAS Traffic management. Unmanned aircraft system traffic management (UTM). https://www.faa.gov/uas/research_development/traffic_management. Accessed 16 June 2020

FAA-H-8083-6 (2021) Advanced Avionics Handbook. Released in 2009, includes all errata/addenda as of 2022. November 2021. FAA, Washington, DC

Fasanella EL, Jackson KE (2001) Crash simulation of a vertical drop test of a B737 fuselage section with auxiliary fuel tank. In: International Aircraft Fire and Cabin Safety Research Conference Proceedings. Atlantic City, NJ, 22–25 Oct 2001. https://www.fire.tc.faa.gov/2001Conference/files/CrashAnalyticalModellingSimulation/EFasanellaPAPER.pdf. Accessed 29 March 2022

Fasanella EL, Carden HD, Boitnott RL, Hayduk RJ (1990) A review of the analytical simulation of aircraft crash dynamics. NASA Technical Memorandum 102595. NASA Langley Research Center, Hampton, VA

Fasanella EL, Jackson KE, Sparks C, Sareen A (2003) Water impact test and simulation of a composite energy absorbing fuselage section. The American Helicopter Society 59th Annual Forum. Phoenix, AZ, 6–8 May 2003. https://ia801206.us.archive.org/7/items/NASA_NTRS_Archive_20030068933/NASA_NTRS_Archive_20030068933.pdf. Accessed 21 May 2021

Fasanella EL, Jackson KE, Kellas S (2008) Soft soil impact testing and simulation of aerospace structures. In: 10th International LS-DYNA Users Conference. Simulation Technology (5), Detroit, MI, pp 18–29, 18–41.

FBI (2020) Federal Bureau of Investigation. Services. Criminal Justice Information Services (CJIS). Biometrics. Next Generation Identification (NGI). https://www.fbi.gov/services/cjis/fingerprints-and-other-biometrics/ngi. Accessed 19 June 2020

Federal Register (2015) Vol. 80, No. 107/Thursday, June 4, 2015/Notices, pp 31946–31948. Department of Transportation. Federal Aviation Administration. Aviation Rulemaking Advisory Committee; Transport airplane and engine issues; New task. https://www.faa.gov/regulations_policies/rulemaking/committees/documents/media/TAEcdeT1-06042015.pdf. Accessed 2 Jan 2020

Fernholz T (2018) Rocket billionaires: Elon Musk, Jeff Bezos, and the new space race. Houghton Mifflin Harcourt, Boston, New York

Fielding JP (1999) Introduction to aircraft design. Cambridge University Press, Cambridge, UK

Forrester JW (1961) Industrial dynamics. Press, Cambridge, MA, The M.I.T

Forrester JW (1969) Urban dynamics. Press, Cambridge, MA, The M.I.T

Fried I (1972) Perturbation errors in the finite element method. J Appl Mech 39(2):629–631

Frings G (1998) FAA Crashworthiness Research Program. International Aircraft Fire and Cabin Safety Research Conference. Atlantic City, NJ, 17 Nov 1998

Fuchs YT, Jackson KE (2008) Vertical drop testing and analysis of the WASP helicopter skid gear. (Presented at The American Helicopter Society 64th Annual Forum, Montreal, Canada, April 29–May 1, 2008). J Am Helicopter Soc 56(1):12005-1–12005-10

Gilbert N, Troitzsch KG (2005) Simulation for the social scientist, 2nd edn. Open University Press, Maidenhead, UK

GUTMA (2017) Global UTM Association. UAS traffic management architecture. April 2017. https://gutma.org/docs/Global_UTM_Architecture_V1.pdf. Accessed 16 June 2020

Golson J (2015) GM's using simulated crashes to build safer cars. WIRED, 9 April 2015. https://www.wired.com/2015/04/gms-using-simulated-crashes-build-safer-cars. Accessed 14 April 2020

Goring G (2016) How to implement an obsolescence plan, utilizing IEC62402 in your business. International Supply Chain Management Conference. Paris, 27 April 2016. https://www.iecq.org/meetings/2016/conference/presentations/IECQ-MC-353-INF_Management_of_Obsolescence.pdf. Accessed 30 March 2022

GTRs (1998) UNECE. Transport. Global Technical Regulations (GTRs). 1998 Agreement on Global Technical Regulations. http://www.unece.org/trans/main/wp29/wp29wgs/wp29gen/wp29glob_registry.html. Accessed 5 April 2020

Guida M, Marulo F, Abrate S (2018) Advances in crash dynamics for aircraft safety. Prog Aerosp Sci 98:106–123

Gumbel E (1962) Statistics of extremes. Columbia University Press, New York

Hall AD (1965) A methodology for systems engineering. D. Van Nostrand Company Inc., Princeton, NJ

Hassler DM, Zeitlin C, Wimmer-Schweingruber RF et al (2013) Mars' surface radiation environment measured with the Mars Science Laboratory's *Curiosity* rover. Science 343(6169)

Hattori S (1997) Energy source for human demand. In: Merz ER and Walter CE (eds) Advanced nuclear systems consuming excess plutonium (NATO ASI Series. 1. Disarmament Technologies—Volume 15), Chapter 8. Springer Science & Business Media, New York, pp 69–78

Haug E, Choi H-Y, Robin S, Beaugonin M (2004) Human models for crash and impact simulation. In: Ciarlet PG (ed). Handbook of numerical analysis, vol. XII. Special volume: Computational models for the human body (Ayache N, guest ed). Elsevier B. V., pp 231–452

Hawkins FH (1993) Human factors in flight, 2nd edn. (Edited by Orlady HW). Ashgate, Aldershot, UK

Heikkilä A-M (1999) Inherent safety in process plant design. An index-based approach. VTT Publications 384. Technical Research Centre of Finland, Espoo

Hilbert M (2014) Scale-free power-laws as interaction between progress and diffusion. Complexity 19(4):56–65

Ho S-Y (ed) (2010) Structural failure analysis and prediction methods for aerospace vehicles and structures. Bentham Science Publishers, Dubai

Holder B (2001) Unmanned air vehicles. An illustrated study of UAVs. Schiffer Publishing Ltd., Atglen, PA

Hollnagel E (2018) Safety-II in practice: developing the resilience potentials. Routledge, London

Hurley TR, Vandenburg JM (eds) (2002) Small airplane crashworthiness design guide. AGATE Reference No. AGATE-WP3.4-034043-036, Simula Technologies Reference No. TR-98099. Simula Technologies, Inc., Phoenix, AZ

Huston RL, Hessel RE, Winget JM (1976) Dynamics of a crash victim—a finite segment model. AIAA J 14(2):173–178

IAEA (1998) International Atomic Energy Agency. Safeguards legal framework—Safeguards agreements. https://www.iaea.org/topics/safeguards-agreements. Accessed 22 Feb 2022

IAEA (2006a) International Atomic Energy Agency. IAEA Safety Standards Series No. SF-1. Fundamental safety principles. Safety fundamentals. IAEA, Vienna

IAEA (2006b) International Atomic Energy Agency. IAEA-TECDOC-1510. Knowledge management for nuclear industry operating organizations. IAEA, Vienna

IAEA (2018) International Atomic Energy Agency. IAEA Safety Standards Series No. SSG-48. Ageing management and development of a programme for long term operation of nuclear power plants. Specific safety guide. IAEA, Vienna

IAEA (2020) International Atomic Energy Agency. IAEA-TECDOC-1919. Application of plant information models to manage design knowledge through the nuclear power plant life cycle. IAEA, Vienna

IAEA (2022) International Atomic Energy Agency. IAEA | PRIS—Power Reactor Information System. The Database on Nuclear Power Reactors, February 2022. https://pris.iaea.org/pris. Accessed 11 Feb 2022

ICAO (1998) International Civil Aviation Organization. Doc 9683. AN/950. Human factors training manual, 1st edn. ICAO, Montréal

ICAO (2003) International Civil Aviation Organization. Doc 9824. AN/450. Human factors guidelines for aircraft maintenance manual. 1st edn. ICAO, Montréal

ICAO (2015) International Civil Aviation Organization. Doc 10019. AN/507. Manual on remotely piloted aircraft systems (RPAS). ICAO, Montréal

ICAO (2017) International Civil Aviation Organization. Annual report 2017. New and emerging activities. Unmanned aerial system Traffic Management (UTM). https://www.icao.int/annual-report-2017/Pages/RU/new-emerging-activities-unmanned-aerial-system-traffic-management.aspx. Accessed 9 Dec 2022

ICAO (2018) International Civil Aviation Organization. Doc 9859. Safety management manual, 4th edn. ICAO, Montréal

IdeaConnection (2015) IdeaConnection by Planbox. Civil aviation safety systems, 17 May 2015. https://www.ideaconnection.com/new-inventions/civil-aviation-safety-systems-09615.html. Accessed 13 May 2022

Illeez M (2015) Why unmanned aircraft systems failed for a century? Int J Sci Eng Res 6(2):451–461

INSAG (1992) International Nuclear Safety Advisory Group. The Chernobyl accident: Updating of INSAG-1. Report INSAG 7. Safety Series No. 75-INSAG-7. IAEA, Vienna

IRIS (2012) International raylway industry standard. Guideline 5: 2012. Obsolescence. UNIFE—The European Rail Industry Association, Brussels

ISAFEM3D (2020) Anwendungsbeispiele für ISAFEM3D. Dynamik: Time history. Raumstation MIR. https://isafem.de/animationen/dynamik/time-history/raumstation-mir. Accessed 2 July 2020

ISO 6385:2016. International standard. Ergonomic principles in the design of work systems. Third edition 2016-08-15. ISO, Geneva (This standard was last reviewed and confirmed in 2021)

ISO/IEC/IEEE 24765:2017. International standard. Systems and software engineering—Vocabulary. ISO, Geneva

Jackson KE, Fasanella EL (1999) Crashworthy evaluation of a 1/5-scale model composite fuselage concept. NASA/TM-1999-209132. ARL-MR-441. NASA Langley Research Center, Hampton, VA. https://ntrs.nasa.gov/api/citations/19990036755/downloads/19990036755.pdf. Accessed 30 March 2022

Jackson KE, Fasanella EL (2001) Crash simulation of a vertical drop test of a B737 fuselage section with overhead bins and luggage. In: The Third Triennial International Aircraft Fire & Cabin Safety Research Conference Proceedings. Atlantic City, NJ, 22–25 Oct 2001. https://www.fire.tc.faa.gov/2001Conference/files/CrashAnalyticalModellingSimulation/KJacksonPAPER.pdf. Also available at NASA Technical Reports Server. https://ntrs.nasa.gov/citations/20040086068. Accessed 30 March 2022

Jackson KE, Fasanella EL (2002) Crash simulation of vertical drop test of two Boeing 737 fuselage sections. Final report. DOT/FAA/AR-02/62. U.S. Army Research Laboratory. NASA Langley Research Center, Hampton, VA. FAA, Washington, DC. https://www.researchgate.net/publication/277864495_Crash_Simulation_of_Vertical_Drop_Tests_of_Two_Boeing_737_Fuselage_Sections. Accessed 29 June 2022

Jackson KE, Fasanella EL (2003) A survey of research performed at NASA Langley Research Center's Impact Dynamics Research Facility. American Institute of Aeronautics and Astronautics, Reston, VA. https://www.researchgate.net/publication/266370987. Accessed 1 Feb 2020

Jackson KE, Boitnott RL, Fasanella EL, Jones LE, Lyle KH (2006) A summary of DOD-sponsored research performed at NASA Langley's impact dynamics research facility. J Am Helicopter Soc 51(1):59–69

Johnson AF, Kindervater CM, Thuis HGSJ, Wiggenraad JFM (1997) Crash resistant composite subfloor structures for helicopters. In: AGARD Conference Proceedings 592 "Advances in Rotorcraft Technology". Advisory Group for Aerospace Research & Development, Neuilly-Sur-Seine, France, 1997, pp. 14–1, 14–12. Papers presented at the Flight Vehicle Integration Panel Symposium held in Ottawa, Canada, 27–30 May 1996

Jones JC (1982) Design methods. John Wiley & Sons, New York

Joseph C (2012) Attention first-class passengers, this is your captain speaking... Crash test proves that it is much safer to sit in the back of the plane. MailOnline, News, 16 September 2012. https://www.dailymail.co.uk/news/article-2203888/The-ultimate-crash-test-dummy-What-passengers-didnt-want-hear-safer-sit-back.html. Accessed 17 Jan 2020

JTSB (2009) Japan transport safety board. Aircraft accident investigation report. China Airlines (Taiwan). Boeing 737-800, B 18616. Spot 41 at Naha Airport. August 20, 2007. Report AA2009-7, Tokyo

Kahneman D, Sibony O, Sunstein CR (2021) Noise: a flaw in human judgment. Little, Brown Spark, New York

Kamal MM (1970) Analysis and simulation of vehicle to barrier impact. In: International Automobile Safety Conference. Technical paper 700714. SAE International, Warrendale, PA

Kamal MM, Lin KH (1982) Collision simulation. In: Kamal MM, Wolf JA (eds) Modern automotive structural analysis. Van Nostrand Reinhold Co., New York, pp 316–355

Kennedy JF (1962) President Kennedy's speech at Rice University, 12 September 1962. YouTube. https://www.youtube.com/watch?v=WZyRbnpGyzQ. Accessed 26 July 2020

King AI (2000) Fundamentals of impact biomechanics: Part I—Biomechanics of the head, neck, and thorax. Annu Rev Biomed Eng 2:55–81

King AI (2001) Fundamentals of impact biomechanics. Part II—Biomechanics of the abdomen, pelvis, and lower extremities. Annu Rev Biomed Eng 3:27–55

Kirkpatrick SW, Schroeder M, Simons JW (2001) Evaluation of passenger rail vehicle crashworthiness. Int J Crashworthiness 6(1):95–106

KNKT (2012) Komite Nasional Keselamatan Transportasi. Final KNKT.12.05.09.04. Aircraft accident investigation report. Sukhoi Civil Aircraft Company. Sukhoi RRJ–95B; 97004. Mount Salak, West Java. Republic of Indonesia. 9 May 2012. National Transportation Safety Committee (KNKT), Ministry of Transportation, Jakarta

KNKT (2019) Komite Nasional Keselamatan Transportasi. Final KNKT.18.10.35.04. Aircraft accident investigation report. PT. Lion Mentari Airlines. Boeing 737-8 (MAX); PK-LQP. Tanjung Karawang, West Java. Republic of Indonesia. 29 October 2018. KNKT, Jakarta

Koenig RL (1996) U.S., Canadian and European aviation authorities coordinate cabin-safety research. Flight Safety Foundation. Cabin Crew Safety, vol. 31, No. 3, May-June 1996. https://flightsafety.org/ccs/ccs_may_june96.pdf. Accessed 30 Jan 2020

Kostek R, Aleksandrowicz P (2017) Simulation of car collision with an impact block. IOP Conf Series: Mater Sci Eng 252:012008

Kotsikos G (2010) Crashworthiness of rail vehicles. NewRail—Centre for Railway Research, Newcastle University, Newcastle, UK. https://docplayer.net/20751203-Crashworthiness-of-rail-vehicles.html. Accessed 15 April 2020

KTA 3201.2 (2017) Sicherheitstechnische Regel des KTA. Komponenten des Primärkreises von Leichtwasserreaktoren. Teil 2: Auslegung, Konstruktion und Berechnung. Fassung 2017-11. Der Kerntechnische Ausschuss (KTA). Salzgitter, Deutschland (Diese Fassung enthält die Berichtigung vom 24. April 2019). http://www.kta-gs.de/d/regeln/3200/3201_1_r_2017_11_ber.pdf. Accessed 31 March 2022

Laananen DH (1974) A digital simulation technique for crashworthy analysis of aircraft seats. Business Aircraft Meeting. Wichita, KS, 2–5 April 1974

Lamanna G, Vanacore A, Guida M, Caputo F, Marulo F, Vitolo B, Cicatiello S (2019) Development of a head injury criteria-compliant aircraft seat by design of experiments. Aerospace 6(95)

Landau E (2017) Voyager 1 fires up thrusters after 37 years. NASA, 1 December 2017. https://www.nasa.gov/feature/jpl/voyager-1-fires-up-thrusters-after-37. Accessed 6 July 2020

Liang X, Fu H, Crosby AJ (2022) Phase-transforming metamaterial with magnetic interactions. Proc Natl Acad Sci 119(1):1–8

Licht DM, Polzella DJ, Boff KR, Armstrong HG (1989) Human factors, ergonomics, and human factors engineering: an analysis of definitions. Technical report. Crew System Ergonomics Information Analysis Centre, Wright Patterson AFB, OH

LSA Code (2017) International life-saving appliance code, 2017th edn. International Maritime Organization, London

Lyle KH, Jackson KE, Fasanella EL (2000) Development of an ACAP helicopter finite element impact model. J Am Helicopter Soc 45(2):137–142

Makhutov NA (2009) Methodology for assessing the risks of terrorism. In: Schweitzer G, Fox M (eds) Russian views on countering terrorism during eight years of dialogue: extracts from proceedings of four U.S.-Russian Workshops. The National Academies Press, Washington, DC, pp 221–235

Mate JL, Silva A (2005) Requirements engineering for sociotechnical systems. Information Science Publishing, Hershey, PA

Matolcsy M (2007) The severity of bus rollover accidents. Society of Mechanical Engineers, Hungary 07, 0989

McComb HG Jr, Thomson RG, Hayduk RJ (1987) Structural dynamics research in a full-scale transport aircraft crash test. J Aircr 24(7):447–453

McEvily AJ (2002) Metal failures: mechanisms, analysis, prevention. John Wiley & Sons, New York

McGuire RJ, Vu T (1998) Vertical drop test of a Beechcraft 1900C airliner. Final report. DOT/FAA/AR-96/119. FAA William J. Hughes Technical Center, Atlantic City International Airport, NJ. U.S. Department of Transportation, FAA, Washington, DC

MGI (2017) McKinsey Global Institute. Artificial intelligence: the next digital frontier? Discussion paper. McKinsey & Company, New York

Mewaldt RA, Davis AJ, Binns WR et al (2005) The cosmic ray radiation dose in interplanetary space—present day and worst-case evaluations. In: 29th International cosmic ray conference, Pune, India, 3–10 August 2005, vol. 00, pp 101–104

MIL-STD-1290A (AV) (1988) Military standard. Light fixed and rotary-wing aircraft crash resistance. U.S. Department of Defense, Washington, DC

Munyazikwiye BB, Vysochinskiy D, Khadyko M, Robbersmyr KG (2018) Prediction of vehicle crashworthiness parameters using piecewise lumped parameters and finite element models. Designs 2(43)

NASA (2020) National Aeronautic and Space Administration. Langley's 2019 Annual Report. Innovation, Discovery, Exploration. NASA Langley Research Center, 13 January 2020. https://www.nasa.gov/langley/overview/2019-annual-report-embed. Accessed 3 Feb 2020

NASA (2021) NASA. Aeronautics Research. UAS Traffic Management (UTM) Project, 30 June 2021. https://www.nasa.gov/utm. Accessed 16 July 2022

NASA Wiki (2016a) National Aeronautics and Space Administration Wiki/Fandom. Controlled Impact Demonstration (CID), 30 August 2016a. https://nasa.fandom.com/wiki/Controlled_Impact_Demonstration. Accessed 6 Feb 2020

NASA Wiki (2016b) National Aeronautics and Space Administration Wiki/Fandom. Launch escape system (LES), 31 August 2016b. https://nasa.fandom.com/wiki/Launch_escape_system. Accessed 23 March 2020

NASA-STD-7009A (2016) NASA Technical Standards. W/ CHANGE 1: Administrative/Editorial Changes. 7 Dec 2016. Standard for models and simulations. https://standards.nasa.gov/sites/default/files/standards/NASA/w/CHANGE-1/1/nasa_std_7009a_change_1.pdf. Accessed 31 March 2022

Newman MEJ (2005) Power laws, Pareto distributions and Zipf's law. Contemp Phys 46(5):323–351

NHTSA (2015) National Highway Traffic Safety Administration. The need for additional heavy truck crashworthiness standards. Report to Congress. U.S. Department for Transportation, NHTSA, Washington, DC

Noland D (2007) What is the safest seat on a plane? Popular Mechanics, 18 July 2007. https://www.popularmechanics.com/flight/a1918/4219452/. Accessed 1 April 2020

NTSB (2015) National Transportation Safety Board. Aerospace accident report. In-flight breakup during test flight, Scaled Composites SpaceShipTwo, N339SS, Near Koehn Dry Lake, California, October 31, 2014. NTSB/AAR-15/02. PB2015-105454. Notation 8614, NTSB, Washington, DC

O'Brien TG, Meister D (2002) Human factors testing and evaluation: an historical perspective. In: Charlton SG, O'Brien TG (eds) Handbook of human factors testing and evaluation, 2nd edn, Chapter 1. Lawrence Erlbaum Associates, Publishers, Mahwah, NJ, pp 5–20

Olivares G (2012) Crashworthiness evaluation of mass transit buses. FTA report No. 0021. National Institute for Aviation Research, Wichita State University, Wichita, KS; U.S. Department for Transportation. Federal Transit Administration, Washington, DC

Oshita F, Omori K, Nakahira Y, Miki K (2002) Development of a finite element model of the human body. In: 7th International LS-DYNA users conference. Crash/Safety (1), Detroit, MI, pp 3–37, 3–48

Ouellette NT, Gordon DM (2021) Goals and limitations of modeling collective behavior in biological systems. Front Phys 9, 687823

Pahl G, Beitz W, Feldhusen J, Grote K-H (2007) Engineering design. A systematic approach, 3rd edn. Springer, London

Peden M, Scurfield R, Sleet D, Mohan D, Hyder AA, Jarawan E, Mathers C (eds) (2004) World report on road traffic injury prevention. World Health Organization, Geneva

Pereira M (2006) Structural crashworthiness of railway vehicles. ResearchGate. https://www.researchgate.net/publication/228843858_Structural_Crashworthiness_of_Railway_Vehicles. Accessed 15 April 2020

PMBOK Guide (2021) A guide to the project management body of knowledge, 7th edn. Project Management Institute, Newtown Square, PA

Presidential Commission (1986) Report of the Presidential Commission on the Space Shuttle Challenger accident. Report to the President. Washington, DC, 6 June 1986. https://history.nasa.gov/rogersrep/genindex.htm. Accessed 1 April 2022

Ralph E (2022) SpaceX launches three Falcon 9 rockets in 36 hours. Teslarati, 19 June 2022. https://www.teslarati.com/spacex-three-falcon-9-rocket-launches-36-hours. Accessed 19 July 2022

Reason J (2000) Human error—models and management. BMJ 320:768–770

Ren Y, Xiang J (2014) Energy absorption structures design of civil aircraft to improve crashworthiness. Aeronaut J New Series 118(1202):383–398

Rettig M (2017) Notes on sociotechnical systems design. Medium, 24 Aug 2017. https://medium.com/rettigs-notes/notes-on-sociotechnical-systems-design-178f161bc9e8. Accessed 22 July 2020

Riccio A, Saputo S, Sellitto A, Russo A, Di Caprio F, Di Palma L (2019) An insight on the crashworthiness behavior of a full-scale composite fuselage section at different impact angles. Aerospace 6(6):72

Roth G, Rebentisch E (2018) The missing sociotechnical systems elements for strategy to implementation. Massachusetts Institute of Technology, Consortium for Engineering Program Excellence (CEPE), 1 Oct 2018. http://cepe.mit.edu/2018/10/01/the-missing-sociotechnical-systems-elements-for-strategy-to-implementation. Accessed 23 July 2020

Ruault J-R, Vanderhaegen F, Luzeaux D (2012) Sociotechnical systems resilience. In: 22nd annual INCOSE international symposium, Rome, Italy, 9–12 July 2012

Russell D, Norvig P (2021) Artificial intelligence: a modern approach, 4th edn. Person, Hoboken, NJ

Saaty TL (1980) The analytic hierarchy process. McGraw Hill, New York

Salvendy G (ed) (1987) Handbook of human factors. John Wiley & Sons, New York

Schneier B (2020) We're banning facial recognition. We're missing the point. The New York Times, 20 January 2020

Schwartz D, Guleyupoglu B, Koya B, Stitzel JD, Gayzik FS (2015) Development of a computationally efficient full human body finite element model. Traffic Inj Prev 16:49–56

Schweitzer G, Fox M (eds) (2009) Russian views on countering terrorism during eight years of dialogue: extracts from proceedings of Four U.S.-Russian Workshops. The National Academies Press, Washington, DC

SEBoK (2021) Guide to the systems engineering body of knowledge, version 2.5. BKCASE, Stevens Institute of Technology, Hoboken, NJ

Seitzberger M, Graf R, Heinzl P, Rittenschober A, Haupt S, Schmidt G (2009) Structural crashworthiness of rail vehicles—from the requirements to the technical solutions. 7th European LS-DYNA Conference. DYNAmore GmbH, Salzburg

Sgobba T, Kanki B, Clervoy J-F, Sandal GM (eds) (2018) Space safety and human performance. Butterworth-Heinemann, Oxford, UK, Cambridge, MA

Shanahan DF (2004) Human tolerance and crash survivability. In: RTO Educational Notes. EN-HFM-113. The Research and Technology Organization of NATO, 2005, pp 6–1, 6–16. Paper presented at the RTO HFM Lecture Series on "Pathological Aspects and Associated Biodynamics in Accident Investigation", held in Madrid, Spain, 28–29 Oct 2004; Königsbrück, Germany, 2–3 Nov 2004

Silver N (2012) The signal and the noise. Why so many predictions fail—but some don't. Random House, New York

Singh H, Ganesan V, Davies J, Paramasuwom M, Gradischnig L (2018) Vehicle interior and restraints modelling development of full vehicle finite element model including vehicle interior and occupant restraints systems for occupant safety analysis using THOR dummies. Report No. DOT HS 812 545. U.S. Department of Transportation, NHTSA, Washington, DC

SOLAS (1974) International convention for the safety of life at sea. Adoption: 1 November 1974; Entry into force: 25 May 1980. International Maritime Organization, London

Spirochkin Y (2013) Hydrodynamic analysis and irrigation device design for the coaxial and bimanual phacoemulsification techniques in cataract surgery. In: Zaidi FH (ed) Cataract surgery, Chater 9. InTech, Rijeka, Croatia, pp 121–136. http://cdn.intechopen.com/pdfs-wm/42693.pdf. Accessed 2 April 2022

Stockwell AE (2002) Simulation of an impact test of the all-composite Lear Fan aircraft. Contractor report. NASA/CR 2002-211458. NASA Langley Research Center, Hampton, VA

Swain AD, Guttmann HE (1983) Handbook of human reliability analysis with emphasis on nuclear power plant applications. NUREG/CR-1278. SAND80-0200. RX, AN. Final report. Sandia National Laboratories, Albuquerque, NM

Synbiosys (2021) Synbiosys. Material science. https://www.synbiosys.co/material-science. Accessed 27 May 2022

Taleb NN (2007). The Black Swan. The impact of the highly improbable. Random House, New York

Tang Z, Liu FJ, Guo SH, Chang J, Zhang JJ (2016) Evaluation of coupled finite element/meshfree method for a robust full-scale crashworthiness simulation of railway vehicles. Adv Mech Eng 8(4):1–13

Torenbeek E (2013) Advanced aircraft design: conceptual design, analysis and optimization of subsonic civil airplanes. John Wiley & Sons, Chichester, UK

Turing AM (1950) Computing machinery and intelligence. Mind 49(236):433–460

Tyrell DC (2002) U.S. rail equipment crashworthiness standards. Proc Inst Mech Eng Part F: J Rail Rapid Transit 216(2):123–130

UN Vehicle Regulations. UNECE. UN Regulations (Addenda to the 1958 Agreement). https://unece.org/un-regulations-addenda-1958-agreement. Accessed 4 April 2020
USAAVSCOM TR 89-D-22A (1989) Aircraft crash survival design guide. Volume I—Design criteria and checklists. Final report. Simula Inc., Phoenix, AZ
USAAVSCOM TR 89-D-22B (1989) Aircraft crash survival design guide. Volume II—Aircraft design crash impact conditions and human tolerance. Final report. Simula Inc., Phoenix, AZ
USAAVSCOM TR 89-D-22C (1989) Aircraft crash survival design guide. Volume III—Aircraft structural crash resistance. Final report. Simula Inc., Phoenix, AZ
USAAVSCOM TR 89-D-22D (1989) Aircraft crash survival design guide. Volume IV—Aircraft seats, restraints, litters, and cockpit/cabin delethalization. Final report. Simula Inc., Phoenix, AZ
USAAVSCOM TR 89-D-22E (1989) Aircraft crash survival design guide. Volume V—Aircraft postcrash survival. Final report. Simula Inc., Phoenix, AZ
USFA (2020) U.S. Fire administration. Emergency services ergonomics and wellness. FA-356. USFA, Emmitsburg, MD
Vangi D, Cialdai C, Gulino M-S, Robbersmyr KG (2018) Vehicle accident databases: correctness checks for accident kinematic data. Designs 2(4)
Vázquez MM (2014) Thoracic injuries in frontal car crashes: risk assessment using a finite element human body model. PhD thesis. Chalmers University of Technology, Gothenburg, Sweden
Vezin P, Verriest JP (2005) Development of a set of numerical human models for safety. In: International Technical Conference on the Enhanced Safety of Vehicles, Washington, DC, June 6–9, 2005. ID (05-0163)
Wang H-P, Pan Y-C, Cheng Y-P (2006) Crashworthiness simulation using coupled mesh-free/finite element formulations in LS-DYNA. In: 9th International LS-DYNA Users Conference. Crash/Safety (4), Dearborn, MI, 4–6 June 2006, pp 16–1, 16–12
Wattles J (2020) NASA-SpaceX mission: astronauts splash down after historic mission. CNN Business, 3 Aug 2020. https://edition.cnn.com/2020/08/02/tech/nasa-spacex-crew-dragon-mission-sunday-scn/index.html. Accessed 4 Aug 2020
Wellkamp P, Meywerk M (2019) Reduction of epistemic uncertainty of a crash box model—experimental and numerical investigations. Latin Am J Solids Struct 16(3):1–18
Wickens CD, Gordon SE, Liu Y (1997) An introduction to human factors engineering. Addison Wesley Longman, New York
Wijker JJ (2008) Spacecraft structures. Springer-Verlag, Berlin
Wirstad J (1979) On the allocation of functions between human and machine. Analysis of a concept and its implication for control system ergonomics. Report No. 13. ERGONOMRÅD AB, Karlstad, Sweden
Wittlin G (1973) A consistent crashworthiness design approach for rotary-wing aircraft. In: 29th Annual National Forum of the American Helicopter Society. Preprint No. 781. Washington, DC
Woodrooffe J, Blower D (2015) Heavy truck crashworthiness: injury mechanisms and countermeasures to improve occupant safety. Report No. DOT HS 812 061. NHTSA, Washington, DC

Sources in Russian

Abrosimov VK, Lebidko VV (2013) Imitatsionnoe modelirovanie organizatsii massovykh meropriyatii (na primere XXII Zimnikh Olimpiiskikh igr 2014) (Simulation of the organization of mass events (on the example of the XXII Winter Olympic Games 2014)). Biznes-informatika (Business-Informatics) 1(23):19–27
Afonsky A, Filipyonok A (2019) Vlasti raskryli detali krusheniya bespilotnika «Orion» v Ryazanskoi oblasti (Authorities revealed details of the crash of the Orion drone in the Ryazan region). RBK. Obschestvo (RBC. Society), 16 Nov 2019. https://www.rbc.ru/society/16/11/2019/5dd018ee9a7947419fc0ed8a. Accessed 26 April 2020

Air Charter Service (2019) Aviatsionnyi Charternyi Servis. O nas. Novosti i materialy. Na chem segodnya letaet rossiiskaya aviatsiya? (Air Charter Service. About us. News and materials. What is Russian aviation flying today?), 1 May 2019. https://www.aircharter.ru/about-us/news-features/blog/aircraft-of-the-russian-aviation. Accessed 27 Aug 2020

AP-25 (2009) Aviatsionnye Pravila. Chast 25. Normy lyotnoi godnosti samolyotov transportnoi kategorii (Aviation rules. Part 25. Airworthiness standards for transport category aircraft), 3rd edn. Interstate Aviation Committee, Moscow

Aptukov AM, Bratsun DA, Lyushnin AV (2013) Modelirovanie povedeniya panikuyuschei tolpy v mnogourovnevom razvetvlennom pomeschenii (Modelling of behavior of panicked crowd in multi-floor branched space). Komp'yuternye Issedovaniya i Modelirovanie (Comput Res Modell) 5(3):491–508

Askarov DT, Bakytzhan DA (2017) Ekonomicheskie aspekty vnedreniya kontseptsii bezlyudnykh proizvodstv (Economic aspects of the introduction of the concept of unmanned production). Nauka bez granits (Science without boundaries) 3(8):16–21

Bashabsheh MM (2014) Kombinirovannaya imitatsionnaya model prostranstvennogo rasprostraneniya epidemii na osnove stokhasicheskoi kompartmentnoi modeli i veroyatnostnogo kletochnogo avtomata (Combined simulation model of spatial distribution of epidemics based on stochastic compartment model and probabilistic cellular automat). Dissertation. Tver State Technical University, Tver

Berestov LM, Kharin EG, Kondratov AA, Miroshnichenko LY, Kalinin YI (2004) Sistema zaschity samolyota ot oshibochnykh ili umyshlennykh deistvii, privodyaschikh k katastrofe (System of protection of aircraft against mishandling or intentional actions resulting in accident). Patent RU 2,228,885 C2, 20 May 2004

Berestov LM, Kharin EG, Yakushev AY, Miroshnichenko LY et al (2010) Avtomatizirovannaya vysokointellektualnaya sistema obespecheniya bezopasnosti polyotov letatelnogo apparata (Automated highly intelligent flight safety system of the aircraft). Patent RU 2,388,663 C1, 10 May 2010

Birbraer AN, Roleder AY (2009) Ekstremalnye vozdeistviya na sooruzheniya (Extreme actions on structures). Publishing House of the Polytechnic University, Saint Petersburg

Bogdanov YV and Timchenko VS (2013) Tekhniko-ekonomicheskie problemy i perspectivy sozdaniya podzemnykh atomnykh stantsii srednei i maloi moschnosti (Technical and economic problems and prospects for the creation of underground nuclear power plants of medium and small power). Nauchno-tekhnicheskie vedomosti Sankt-Peterburgskogo gosudarstvennogo politekhnicheskogo universiteta. Energetika. Elektrotekhnika (Scientific and technical bulletin of Saint Petersburg State Polytechnic University. Energy. Electr Eng) 3(178):71–81

Boiko OG (2010) Sovershenstvovanie metodov raschyota nadezhnosti funktsionalnykh sistem samolyotov grazhdanskoi aviatsii i issledovanie protsessov stareniya (Improvement of methods for analyzing the reliability of functional systems of civil aircraft and the study of aging processes). Dissertation. Academician M. F. Reshetnev Siberian State Aerospace University, Krasnoyarsk

Bolotin VV (1971) Primenenie metodov teorii veroyatnostei i teorii nadezhnosti v raschyotakh sooruzhenii (Application of the methods of theory of probability and theory of reliability to analysis of structures). Publishing house for literature on construction, Moscow

Breer VV, Novikov DA, Rogatkin AD (2016) Upravlenie tolpoi: matematicheskie modeli porogovogo kollektivnogo povedeniya (Crowd management: mathematical models of threshold collective behavior). Lenard, Moscow (Smart Management Series)

Chumak DY, Schepetina TD (2015) Riski atomno-energeticheskikh proektov: podkhody k klassifikatsii i upravleniyu. V kn.: Atomnye stantsii maloi moschnosti: novoe napravlenie razvitiya energetiki. Tom 2. Pod red. akad. RAN A. A. Sarkisova (Risks of nuclear power projects: approaches to classification and management. In: Sarkisov AA (ed) Low-power nuclear power plants: a new line in the development of power systems, vol 2. Academ-Print, Moscow, pp 299–314

Daveze F, Foucart V, Botargues P, Averseng D (2009) Sposob i sistema dlya ukloneniya ot slolknovenii dlya letatelnogo apparata (Avoidance method and system for an aircraft). Patent RU 2,343,528 C1, 10 Jan 2009

Dolgov VN (2006) Podzemnaya atomnaya elektrostantsiya (Underground nuclear power plant). Patent RU 2,273,901 C2, 10 April 2006

Dorokhova I (2020) Iskusstvo rabotat s intellektom (The art of working with intelligence). Strana Rosatom (Country Rosatom) No. 2(418):8–9

Dynamika (1992) Analiz regionalnoi bezopasnosti territorii Rosiiskoi Federatsii s uchetom riska padeniya aviatsionnykh i kosmicheskikh ob'ektov. Itogovyi otchet po dogovoru № 12–92 s Goskomitetom RF po chrezvychainym situatsiyam (Analysis on regional safety of the territory of the Russian Federation taking into account the risk of falling aviation and space objects. Final report on the agreement No. 12–92 with the State Committee of the Russian Federation for Emergency Situations). Research and engineering company *Dynamika*, Korolyov, Moscow region

ENLG (1985) Edinye Normy Lyotnoi Godnosti grazhdanskikh transportnykh samolyotov stran—chlenov SEV (Uniform airworthiness standards for civil transport aircraft of the Comecon member states). TsAGI, Moscow

GOST 2.102-2013 Mezhgosudarstvennyi standart. Edinaya sistema konstruktorskoi dokumentatsii. Vidy I komplektnost konstruktorskikh dokumentov (Interstate Standard. Unified system for design documentation. Types and sets of design documentation). Standartinform, Moscow, 2014

GOST 2.103-2013 Mezhgosudarstvennyi standart. Edinaya sistema konstruktorskoi dokumentatsii. Stadii razrabotki (Interstate Standard. Unified system for design documentation. Stages of designing). Standartinform, Moscow, 2015

GOST 2.118-2013 Mezhgosudarstvennyi standart. Edinaya sistema konstruktorskoi dokumentatsii. Tekhnicheskoe predlozhenie (Interstate Standard. Unified system for design documentation. Technical proposal). Standartinform, Moscow, 2018

GOST 2.119-2013 Mezhgosudarstvennyi standart. Edinaya sistema konstruktorskoi dokumentatsii. Eskiznyi proekt (Interstate Standard. Unified system for design documentation. Preliminary design). Standartinform, Moscow, 2018

GOST 2.120-2013 Mezhgosudarstvennyi standart. Edinaya sistema konstruktorskoi dokumentatsii. Tekhnicheskii proekt (Interstate Standard. Unified system for design documentation. Technical design). Standartinform, Moscow, 2015

GOST 29.05.002-82 Gosudarstvennyi standart Soyuza SSR. Sistema standartov ergonomicheskikh trebovanii i ergonomicheskogo obespecheniya. Indikatory tsifrovye znakosinteziruyuschie. Obschie ergonomicheskie trebovaniya (State Standard of the USSR. Standards system of ergonomic requirements and ergonomic means. Digital and sign-synthesizing indicators. General ergonomic requirements). State Committee of the USSR for Standards, Moscow, 1983

GOST 20296-2014 Mezhgosudarstvennyi standart. Samolyoty i vertolyoty grazhdanskoi aviatsii. Dopustimye urovni shuma v salonakh i kabinakh ekipazha i metody izmereniya shuma (Interstate Standard. Airplanes and helicopters of civil aviation. Acceptable noise levels in flight decks and in salons and methods of noise measurement). Standartinform, Moscow, 2014

GOST 21958-76 Gosudarstvennyi standart Soyuza SSR. Systema «Chelovek-mashina». Zal i kabiny operatorov. Vzaimnoe raspolozhenie rabochikh mest. Obschie ergonomicheskie trebovaniya (State Standard of the USSR. Man-machine system. Operator's rooms and cabin. General ergonomics requirements). State Committee of Standards of the Council of Ministers of the USSR, Moscow, 1977

GOST 22269-76 Gosudarstvennyi standart Soyuza SSR. Systema «Chelovek-mashina». Rabochee mesto operatora. Vzaimnoe raspolozhenie elementov rabochego mesta. Obschie ergonomicheskie trebovaniya (State Standard of the USSR. Man-machine system. Operator's workplace. Arrangement of workplace elements. General ergonomic requirements). Standards Publishing House, Moscow, 1990

GOST 23000-78 Gosudarstvennyi standart Soyuza SSR. Systema «Chelovek-mashina». Pulty upravleniya. Obschie ergonomicheskie trebovaniya (State Standard of the USSR. Man-machine system. Control consoles. General ergonomic requirements). State Committee of Standards of the Council of Ministers of the USSR, Moscow, 1979

GOST 32410-2013 (EN 15227:2008 + A1:2010, NEQ) Mezhgosudarstvennyi standart. Krash-sistemy avariinye zheleznodorozhnogo podvizhnogo sostava dlya passazhirskikh perevozok. Tekhnicheskie trebovaniya i metody kontrolya (Interstate Standard. Emergency crash-systems of railway rolling stock for passenger transportations. Technical requirements and methods of control). Standartinform, Moscow, 2014

GOST 33128-2014 Mezhgosudarstvennyi standart. Dorogi avtomobilnye obschego polzovaniya. Ograzhdeniya dorozhnye. Tekhnicheskie trebovaniya (Interstate Standard. Automobile roads of general use. Road restraint systems. Technical requirements). Standartinform, Moscow, 2019

GOST 33129-2014 Mezhgosudarstvennyi standart. Dorogi avtomobilnye obschego polzovaniya. Ograzhdeniya dorozhnye. Metody kontrolya (Interstate Standard. Automobile roads of general use. Road restraint systems. Methods of testing). Standartinform, Moscow, 2015

GOST ISO 9000-2011 (ISO 9000:2005, IDT) Mezhgosudarstvennyi standart. Sistemy menedzhmenta kachestva. Osnovnye polozheniya i slovar (Interstate Standard. Quality management systems. Fundamentals and vocabulary). Standartinform, Moscow, 2012

GOST ISO 9001-2011 (ISO 9001:2008, IDT) Mezhgosudarstvennyi standart. Sistemy menedzhmenta kachestva. Trebovaniya (Interstate Standard. Quality management systems. Requirements). Standartinform, Moscow, 2018

GOST R 2.601-2019 Natsionalnyi standart Rossiiskoi Federatsii. Edinaya sistema konstruktorskoi dokumentatsii. Ekspluatatsionnye dokumenty (National Standard of the Russian Federation. Unified system for design documentation. Exploitative documents). Standartinform, Moscow, 2019

GOST R 10.0.02-2019 (ISO 16739-1:2018, IDT) Natsionalnyi standart Rossiiskoi Federatsii. Sistema standartov informatsionnogo modelirovaniya zdanii i sooruzhenii. Otraslevye bazovye klassy (IFC) dlya obmena i upravleniya dannymi ob ob'ektakh stroitelstva. Chast 1. Skhema dannykh (National Standard of the Russian Federation. System of standards on information modeling of buildings and structures. Industry Foundation Classes (IFC) for data sharing in the construction and facility management industries. Part 1. Data schema). Standartinform, Moscow, 2019

GOST R 10.0.03-2019 (ISO 29481-1:2016, IDT) Natsionalnyi standart Rossiiskoi Federatsii. Sistema standartov informatsionnogo modelirovaniya zdanii i sooruzhenii. Informatsionnoe modelirovanie v stroitelstve. Spravochnik po obmenu informatsiei. Chast 1. Metodologiya i format (National Standard of the Russian Federation. System of standards on information modeling of buildings and structures. Building information models. Information delivery manual. Part 1. Methodology and format). Standartinform, Moscow, 2019

GOST R 10.0.04-2019 (ISO 29481-2:2012, IDT) Natsionalnyi standart Rossiiskoi Federatsii. Sistema standartov informatsionnogo modelirovaniya zdanii i sooruzhenii. Informatsionnoe modelirovanie v stroitelstve. Spravochnik po obmenu informatsiei. Chast 2. Struktura vzaimodeistviya (National Standard of the Russian Federation. System of standards on information modeling of buildings and structures. Building information models. Information delivery manual. Part 2. Interaction framework). Standartinform, Moscow, 2019

GOST R 10.0.05-2019 (ISO 12006-2:2015, IDT) Natsionalnyi standart Rossiiskoi Federatsii. Sistema standartov informatsionnogo modelirovaniya zdanii i sooruzhenii. Stroitelstvo zdanii. Struktura informatsii ob ob'ektakh stroitelstva. Chast 2. Osnovnye printsipy klassifikatsii (National Standard of the Russian Federation. System of standards on information modeling of buildings and structures. Building construction. Organization of information about construction works. Part 2. Framework for classification). Standartinform, Moscow, 2019

GOST R 10.0.06-2019 (ISO 12006-3:2007, IDT) Natsionalnyi standart Rossiiskoi Federatsii. Sistema standartov informatsionnogo modelirovaniya zdanii i sooruzhenii. Stroitelstvo zdanii. Struktura informatsii ob ob'ektakh stroitelstva. Chast 3. Osnovy obmena ob'ektno-orientirovannoi informatsiei (National Standard of the Russian Federation. System of standards on information modeling of buildings and structures. Building construction. Organization of information about construction works. Part 3. Framework for object-oriented information). Standartinform, Moscow, 2019

GOST R 12.0.001-2013 Natsionalnyi standart Rossiiskoi Federatsii. Sistema standartov bezopasnosti truda. Obschie polozheniya (National Standard of the Russian Federation. Occupational safety standards system. Basic rules). Standartinform, Moscow, 2019

GOST R 22.0.02-2016 Natsionalnyi standart Rossiiskoi Federatsii. Bezopasnost v chrezvychainykh situatsiyakh. Terminy i opredeleniya (National Standard of the Russian Federation. Safety in emergencies. Terms and definitions). Standartinform, Moscow, 2016

GOST R 22.0.03-95 Gosudarstvennyi standart Rossiiskoi Federatsii. Bezopasnost v chrezvychainykh situatsiyakh. Prirodnye chrezvychainye situatsii. Terminy i opredeleniya (State Standard of the Russian Federation. Safety in emergencies. Natural emergencies. Terms and definitions). Standards Publishing House, Moscow, 1995

GOST R 22.3.01-94 Gosudarstvennyi standart Rossiiskoi Federatsii. Bezopasnost v chrezvychainykh situatsiyakh. Zhizneobespechenie naseleniya v chrezvychainykh situatsiyakh. Obschie trebovaniya (State Standard of the Russian Federation. Safety in emergencies. Human life support in emergencies. General requirements). Standards Publishing House, Moscow, 1995

GOST R 29.05.008-96 Gosudarstvennyi standart Rossiiskoi Federatsii. Sistema standartov ergonomicheskikh trebovanii i ergonomicheskogo obespecheniya. Rabochee mesto dispetchera sluzhb upravleniya vozdushnym dvizheniem. Obschie ergonomicheskie trebovaniya (State Standard of the Russian Federation. System of standards for ergonomic requirements and ergonomic support. Workplace of air traffic control services dispatcher. General ergonomic requirements). Gosstandart of Russia, Moscow, 1996

GOST R 29.08.004-96 Gosudarstvennyi standart Rossiiskoi Federatsii. Sistema standartov ergonomicheskikh trebovanii i ergonomicheskogo obespecheniya. Rabochee mesto dispetchera sluzhb upravleniya vozdushnym dvizheniem. Metody otsenki sootvetstviya obschim ergonomicheskim trebovaniyam (State Standard of the Russian Federation. System of standards for ergonomic requirements and ergonomic support. Workplace of air traffic control services dispatcher. Methods for assessing compliance with general ergonomic requirements). Gosstandart of Russia, Moscow, 1996

GOST R 51897-2021 (ISO Guide 73:2009) Natsionalnyi standart Rossiiskoi Federatsii. Menedzhment riska. Terminy i opredeleniya (National Standard of the Russian Federation (ISO Guide 73:2009). Risk management—Vocabulary, MOD). Rossiiskii institut standartizatsii, Moscow, 2021

GOST R 51901.1-2002 Gosudarstvennyi standart Rossiiskoi Federatsii. Menedzhment riska. Analiz riska tekhnologicheskikh sistem (State Standard of the Russian Federation. Risk management. Risk analysis of technological systems). Standards Publishing House, Moscow, 2002

GOST R 51901.23-2012 Natsionalnyi standart Rossiiskoi Federatsii. Menedzhment riska. Reestr riska. Rukovodstvo po otsenke riska opasnykh sobytii dlya vklyucheniya v reestr riska (National Standard of the Russian Federation. Risk management. Risk register. Guide on assessment of hazards risk for inclusion in risk register). Standartinform, Moscow, 2014

GOST R 54088-2017 Natsionalnyi standart Rossiiskoi Federatsii. Integrirovannaya logisticheskaya podderzhka. Ekspluatatsionnaya i remontnaya dokumentatsiya v forme interactivnykh elektronnykh tekhnicheskikh rukovodstv. Osnovnye polozheniya i obschie trebovaniya (National Standard of the Russian Federation. Integrated logistic support. Operating and maintenance documentation in interactive electronic technical manuals format. General provisions and general requirements). Standartinform, Moscow, 2018

GOST R 54869-2011 Natsionalnyi standart Rossiiskoi Federatsii. Proektnyi menedzhment. Trebovaniya k upravleniyu proektom (National Standard of the Russian Federation. Project management. Requirements for project management). Standartinform, Moscow, 2012

GOST R 55237.1-2012 Natsionalnyi standart Rossiiskoi Federatsii. Ergonomika transportnykh sredstv. Ergonomicheskie aspekty informatsionno-upravlyayuschei sistemy transportnogo sredstva. Protsedury opredeleniya prioriteta soobschenii voditelyu (National Standard of the Russian Federation. Ergonomics of vehicles. Ergonomic aspects of transport information and control system. Procedures for determining priority of on-board messages presented to drivers). Standartinform, Moscow, 2018

GOST R 55241.1-2012 Natsionalnyi standart Rossiiskoi Federatsii. Ergonomika vzaimodeistviya chelovek-sistema. Chast 100. Vvedenie v standarty, otnosyaschiesya k ergonomike programmnykh sredstv (National Standard of the Russian Federation. Ergonomics of human-system interaction. Part 100. Introduction to standards related to software ergonomics). Standartinform, Moscow, 2013

GOST R 55249-2012 Natsionalnyi standart Rossiiskoi Federatsii. Vozdushnyi transport. Aeroporty. Tekhnicheskie sredtva dosmotra. Obschie tekhnicheskie trebovaniya (National Standard of the Russian Federation. Air transport. Airports. Technical means for screening. General technical requirements). Standartinform, Moscow, 2013

GOST R 56129-2014 (MEK 62402:2007) Natsionalnyi standart Rossiiskoi Federatsii. Integrirovannaya logisticheskaya podderzhka eksportiruemoi produktsii voennogo naznacheniya. Upravlenie nomenklaturoi ustarevayuschikh pokupnykh komplektuyuschikh izdelii (National Standard of the Russian Federation. Integrated logistic support of exporting military products. Obsolescence management). Standartinform, Moscow, 2015

GOST R 56136-2014 Natsionalnyi standart Rossiiskoi Federatsii. Upravlenie zhiznennym tsiklom produktsii voennogo naznacheniya. Terminy i opredeleniya (National Standard of the Russian Federation. Life cycle management for military products. Terms and definitions). Standartinform, Moscow, 2015

GOST R 56274-2014 Natsionalnyi standart Rossiiskoi Federatsii. Obschie pokazateli i trebovaniya v ergonomike (National Standard of the Russian Federation. General ergonomics requirements and properties). Standartinform, Moscow, 2015

GOST R 56620.2-2015/ISO/TR 7250-2:2010 Natsionalnyi standart Rossiiskoi Federatsii. Ergonomika. Osnovnye antropometricheskie izmereniya dlya tekhnicheskogo proektirovaniya. Chast 2. Statisticheskie dannye natsionalnykh sovokupnostei (National Standard of the Russian Federation. Ergonomics. Basic anthropometrical measurements for technical designing. Part 2. Statistical data of national populations). Standartinform, Moscow, 2016

GOST R 57188-2016 Natsionalnyi standart Rossiiskoi Federatsii. Chislennoe modelirovanie fizicheskikh protsessov. Terminy i opredeleniya (National Standard of the Russian Federation. Numerical modeling of physical processes. Terms and definitions). Standartinform, Moscow, 2016

GOST R 57193-2016 (ISO/IEC/IEEE 15288:2015, NEQ) Natsionalnyi standart Rossiiskoi Federatsii. Sistemnaya i programmnaya inzheneriya. Protsessy zhiznennogo tsikla sistem (National Standard of the Russian Federation. Systems and software engineering. System life cycle processes). Standartinform, Moscow, 2016

GOST R 57700.1-2017 Natsionalnyi standart Rossiiskoi Federatsii. Chislennoe modelirovanie dlya razrabotki i sdachi v ekspluatatsiyu vysokotekhnologichnykh promyshlennykh izdelii. Sertifikatsiya programmnogo obespecheniya. Trebovaniya (National Standard of the Russian Federation. Numerical simulation for the development and commissioning of high-tech industrial products. Software certification. Requirements). Standartinform, Moscow, 2018

GOST R 57700.2-2017 Natsionalnyi standart Rossiiskoi Federatsii. Chislennoe modelirovanie dlya razrabotki i sdachi v ekspluatatsiyu vysokotekhnologichnykh promyshlennykh izdelii. Sertifikatsiya programmnogo obespecheniya. Obschie polozheniya (National Standard of the Russian Federation. Numerical simulation for the development and commissioning of high-tech industrial products. Software certification. General provisions). Standartinform, Moscow, 2017

GOST R 57700.3-2017 Natsionalnyi standart Rossiiskoi Federatsii. Chislennoe modelirovanie dynamicheskikh rabochikh protsessov v sotsiotekhnicheskikh sistemakh. Terminy i opredeleniya (National Standard of the Russian Federation. Numerical modeling of dynamic work processes in sociotechnical systems. Terms and definitions). Standartinform, Moscow, 2018

GOST R 57700.4-2017 Natsionalnyi standart Rossiiskoi Federatsii. Chislennoe modelirovanie fizicheskikh protsessov. Terminy i opredeleniya v oblastyakh mekhaniki sploshnykh sred: gidromekhanika, gazovaya dinamika (National Standard of the Russian Federation. Numerical modeling of physical processes. Terms and definitions in the fields of continuum mechanics: fluid mechanics, gas dynamics). Standartinform, Moscow, 2018

GOST R 57700.6-2017 Natsionalnyi standart Rossiiskoi Federatsii. Chislennoe modelirovanie fizicheskikh protsessov. Terminy i opredeleniya v oblasti bessetochnykh metodov chislennogo modelirovaniya (National Standard of the Russian Federation. Numerical modeling of physical processes. Terms and definitions for numerical meshfree methods). Standartinform, Moscow, 2017

GOST R 57700.7-2017 Natsionalnyi standart Rossiiskoi Federatsii. Chislennoe modelirovanie fizicheskikh protsessov. Protsessy udarnogo vzaimodeistviya. Terminy i opredeleniya (National Standard of the Russian Federation. Numerical modeling of physical processes. Processes of impact interaction. Terms and definitions). Standartinform, Moscow, 2018

GOST R 58217-2018 Natsionalnyi standart Rossiiskoi Federatsii. Neftyanaya i gazovaya promyshlennost. Arkticheskie operatsii. Evakuatsiya i spasanie personala morskikh platform. Obschie polozheniya (National Standard of the Russian Federation. Petroleum and natural gas industries. Arctic operations. Evacuation and rescue of offshore platform personnel. General provisions). Standartinform, Moscow, 2018

GOST R 58776-2019 Natsionalnyi standart Rossiiskoi Federatsii. Sredstva monitoringa povedeniya i prognozirovaniya namerenii lyudei. Terminy i opredeleniya (National Standard of the Russian Federation. Tools for monitoring behavior and predicting people's intentions. Terms and definitions). Standartinform, Moscow, 2020

GOST R 58777-2019 Natsionalnyi standart Rossiiskoi Federatsii. Vozdushnyi transport. Aeroporty. Tekhnicheskie sredstva dosmotra. Metodika opredeleniya pokazatelei kachestva raspoznavaniya nezakonnykh vlozhenii po tenevym rentgenovskim izobrazheniyam (National Standard of the Russian Federation. Air transport. Airports. Technical means of inspection. Methodology for determination of quality indicators of recognition of illegal attachments by shadow X-ray images). Standartinform, Moscow, 2020

GOST R ISO 6385-2016 (ISO 6385:2016, IDT) Natsionalnyi standart Rossiiskoi Federatsii. Ergonomika. Primenenie ergonomicheskikh printsipov pri proektirovanii proizvodstvennykh sistem (National Standard of the Russian Federation. Ergonomics. Ergonomic principles in the design of work systems). Standartinform, Moscow, 2016

GOST R ISO 7250-1-2013 (ISO 7250-1:2008, IDT) Natsionalnyi standart Rossiiskoi Federatsii. Ergonomika. Osnovnye antropometricheskie izmereniya dlya tekhnicheskogo proektirovaniya. Chast 1. Opredeleniya i osnovnye antropometricheskie tochki (National Standard of the Russian Federation. Ergonomics. Basic human body measurements for technological design. Part 1. Body measurement definitions and landmarks). Standartinform, Moscow, 2014

GOST R ISO 7250-3-2019 (ISO 7250-3:2015, IDT) Natsionalnyi standart Rossiiskoi Federatsii. Ergonomika. Osnovnye antropometricheskie izmereniya dlya tekhnicheskogo proektirovaniya. Chast 3. Mezhdunarodnye i regionalnye dannye dlya ispolzovaniya v standartakh na produktsiyu (National Standard of the Russian Federation. Ergonomics. Basic human body measurements for technological design. Part 3. Worldwide and regional design ranges for use in product standards). Standartinform, Moscow, 2019.

GOST R ISO 9241-161-2016 (ISO 9241-161:2016, IDT) Natsionalnyi standart Rossiiskoi Federatsii. Ergonomika vzaimodeistviya chelovek-sistema. Chast 161. Elementy graficheskogo polzovatelskogo interfeisa (National Standard of the Russian Federation. Ergonomics of human-system interaction. Part 161. Visual user-interface elements). Standartinform, Moscow, 2016

GOST R ISO 10006-2005 (ISO 10006:2003, IDT) Natsionalnyi standart Rossiiskoi Federatsii. Sistemy menedzhmenta kachestva. Rukovodstvo po menedzhmentu kachestva pri proektirovanii (National Standard of the Russian Federation. Quality management systems. Guidelines for quality management in projects). Standartinform, Moscow, 2007

GOST R ISO 11064-1-2015 (ISO 11064-1:2000, IDT) Natsionalnyi standart Rossiiskoi Federatsii. Ergonomicheskoe proektirovanie tsentrov upravleniya. Chast 1. Printsipy proektirovaniya (National Standard of the Russian Federation. Ergonomic design of control centres. Part 1. Principles for the design). Standartinform, Moscow, 2016

GOST R ISO 11064-2-2015 (ISO 11064-2:2000, IDT) Natsionalnyi standart Rossiiskoi Federatsii. Ergonomicheskoe proektirovanie tsentrov upravleniya. Chast 2. Printsipy organizatsii kompleksov upravleniya (National Standard of the Russian Federation. Ergonomic design of control centres. Part 2. Principles for the arrangement of control suites). Standartinform, Moscow, 2016

GOST R ISO 15926-1-2008 (ISO 15926-1:2004, IDT) Natsionalnyi standart Rossiiskoi Federatsii. Promyshlennye avtomatizirovannye sistemy i integratsiya. Integratsiya dannykh zhiznennogo tsikla dlya pererabatyvayuschikh predpriyatii, vklyuchaya neftyanye i gazovye proizvodstvennye predpriyatiya. Chast 1. Obzor i osnovopolagayuschie printsipy (National Standard of the Russian Federation. Industrial automation systems and integration. Integration of life-cycle data for process plants including oil and gas production facilities. Part 1. Overview and fundamental principles). Standartinform, Moscow, 2019

GOST R ISO 15926-2-2010 (ISO 15926-2:2003, IDT) Natsionalnyi standart Rossiiskoi Federatsii. Sistemy promyshlennoi avtomatizatsii i integratsiya. Integratsiya dannykh zhiznennogo tsikla dlya pererabatyvayuschikh predpriyatii, vklyuchaya neftyanye i gazovye proizvodstvennye predpriyatiya. Chast 2. Model dannykh (National Standard of the Russian Federation. Industrial automation systems and integration. Integration of life-cycle data for process plants including oil and gas production facilities. Part 2. Data model) Standartinform, Moscow, 2013

GOST R ISO 21500-2014 (ISO 21500:2012, IDT) Natsionalnyi standart Rossiiskoi Federatsii. Rukovodstvo po proektnomu menedzhmentu (National Standard of the Russian Federation. Guidance on project management). Standartinform, Moscow, 2015

GOST R ISO 26800-2013 (ISO 26800:2011, IDT) Natsionalnyi standart Rossiiskoi Federatsii. Ergonomika. Obschie printsipy i ponyatiya (National Standard of the Russian Federation. Ergonomics. General principles and concepts). Standartinform, Moscow, 2014

Gusarov VV, Almyashev VI, Khabenskii VB, Beshta SV, Granovskii VS (2005) Novyi klass funktsionalnykh materialov dlya ustroistva lokalizatsii rasplava aktivnoi zony yadernogo reaktora (A new class of functional materials for the core melt localization device of a nuclear reactor). Rossiiskii khimicheskii zhurnal (Zhurnal Rossiiskogo khimicheskogo obschestva im. D. I. Mendeleeva) (Russian Chem J D. I. Mendeleev Russian Chem Soc) XLIX(4):42–53

Il'enkova SD, Bandurin AV, Gorbovtsev GY et al (2002) Proizvodstvennyi menedzhment: Uchebnik dlya vuzov. Pod red. S. D. Il'enkovoi (Production management: Textbook for universities. Edited by S. D. Il'enkova). YuNITI-DANA, Moscow

IMASh (1992) Institut mashinovedeniya im. A. A. Blagonravova Rossiiskoi Akademii nauk (IMASh). Gosudarstvennaya nauchno-tekhnicheskaya programma «Bezopasnost». Razrabotka printsipov i kriteriev normalnykh uslovii raboty, otklonenii ot normalnykh uslovii, avariinykh situatsii, proektnykh, zaproektnykh i gipoteticheskikh avarii. Nauch. ruk. proekta N. A. Makhutov. Otchet o nauchno-issledovatelskoi rabote po proektu 1.3 (A. A. Blagonravov Mechanical Engineering Research Institute of the Russian Academy of Sciences (IMASh). State scientific and technical program "Safety". Development of principles and criteria for normal operation conditions, deviations from normal conditions, emergencies, design basis accidents, beyond-design-basis accidents and hypothetical accidents. Project scientific leader N. A. Makhutov. Report on the research work on the project 1.3). IMASh, Moscow (Contributors: Bugaenko SE, Bykov DL, Karmishin AV, Kutepov SM, Mannapov RG, Spirochkin YK et al)

Ipatova Y (2019) Era termoyadernogo sinteza (The era of thermonuclear fusion). Atomnyi ekspert. Tekhnologii (Atomic expert. Technologies) #2-3_2019. http://atomicexpert.com/era_of_thermonuclear_fusion. Accessed 9 March 2020

Karpov YG (2005) Imitatsionnoe modelirovanie sistem. Vvedenie v modelirovanie s AnyLogic 5 (Simulation modelling of systems. Introduction to modelling with AnyLogic 5). BHV-Peterburg, Saint Petersburg

Khusainov AS, Kuzmin YA (2011) Passivnaya bezopasnost avtomobilya: Uchebnoe posobie (Passive car safety: Tutorial). Ul'yanovsk State Technical University, Ul'yanovsk

Klyuev VV (ed) (1982) Ispytatelnaya tekhnika: Spravochnik. V 2 kn. Pod red. Klyueva VV. Kn. 1 (Test engineering: reference manual. In 2 books. Book 1). Mashinostroenie, Moscow

Klyuev VV (ed) (2003) Mashinostroenie. Entsiklopediya. Nadezhnost mashin. Tom IV-3. Klyuev VV, Bolotin FR, Sosnin i dr. Pod obsch. red. Klyueva VV (Mechanical engineering. Encyclopedia.

Reliability of machines. Volume IV-3. Contributors: Klyuev VV, Bolotin VV, Sosnin FR et al). Mashinostroenie, Moscow

Kolpaksidi AP (2020) Kosmonavtika v 2019 godu: statistika, dinamika, vazhneishie sobytiya (Cosmonautics in 2019: statistics, dynamics, the most important events). Yandex Dzen. Zemlya i Vselennaya (Yandex Zen. Earth and the Universe), 5 Jan 2020. https://zen.yandex.ru/media/id/5acb60ccdb0cd9b106504c84/kosmonavtika-v-2019-godu-statistika-dinamika-vajneishie-sobytiia-5e0b49ac3f548700ad63a374?utm_source=serp. Accessed 4 Aug 2020

Komarov AA (1969) Nadezhnost gidravlicheskikh sistem (Reliability of hydraulic systems). Mashinostroenie, Moscow

Krylov AN (2016) Moi vospominaniya. Izd. 9, ispr. i dop. (My memories, 9th edn, revised and supplemented). LENARD, Moscow

Kurpatov AV (2018) Neironauka o narastayuschei gluposti (Neuroscience about increasing stupidity). Fontanka.Ru. Saint Petersburg on line, 15.01.2018. https://www.fontanka.ru/2018/01/15/060. Accessed 17 July 2020

Kurpatov AV (2020) Vystuplenie v Davose na biznes-zavtrake Sberbanka (Speech at the Sberbank business breakfast in Davos). YouTube.Ru, 29 Jan 2020. https://www.youtube.com/watch?v=GbLclnu-QGc. Accessed 21 Feb 2020

Lebon G (2017) Psikhologiya narodov i mass (Psychology of peoples and masses). Publishing House AST, Moscow. Translated into Russian from French original: Gustave Le Bon. Lois psychologiques de l'évolution des peuples (1894) & La psychologie des foules (1895)

Lem S (2019) Summa tekhnologii (Summa technologiae). Publishing House AST, Moscow. Translated into Russian from Polish original: Stanisław Lem. Summa Technologiae (1967)

Lyakhovenko IA, Merkur'ev AV (1995) Raschyot nagruzok v sisteme «kreslo-chelovek-privyaznye remni» pri deistvii na letatelnyi apparat peregruzok v avariinykh situatsiyakh (Calculation of loads in the seat-human-belts system when loading an aircraft due to accelerations in emergency situations). Uchenye zapiski TsAGI (Scientific notes of TsAGI) XXVI(3–4):138–146

Lyakhovenko IA, Shenk YV, Spirochkin YK (1989) K voprosu o vozmozhnosti chislennogo modelorovaniya avariinoi posadki vertolyota. Nauchno-tekhnicheskii otchet NTO 03-5942 (On the possibility of numerical simulation of a helicopter emergency landing. Scientific and technical report NTO 03-5942). TsAGI, Zhukovsky, Moscow region

MAK (2013) Mezhgosudarstvennyi Aviatsionnyi Komitet. Okonchatelnyi otchet po rezultatam rassledovaniya aviatsionnogo proisshestviya: Boeing 737-500 VQ-BBN (Interstate Aviation Committee. Final report on the results of the accident investigation: Boeing 737-500 VQ-BBN). IAC, Moscow

Malikova DM (2017) Modelirovanie effektivnogo mekhanizma programmno-proektnogo upravleniya vnedreniem bezlyudnykh tekhnologii proizvodstva na predpriyatiyakh OPK (Modeling of an effective mechanism for program-project management of the introduction of unmanned production technologies at defense industry enterprises). Organizator proizvodstva (Prod Organizer) 25(4):18–31

Moiseev NN (1979) Matematika stavit eksperiment (Mathematics sets up an experiment). Nauka, Moscow

Nesterenko BG, Nesterenko GI (2014) Resurs konstruktsii stareyuschikh transportnykh samolyotov (The life of structures of aging transport airplanes). Nauchnyi vestnik MGTUGA (Moskovskogo gosudarstvennogo tekhnicheskogo universiteta grazhdanskoi aviatsii) (Scientific Bulletin of MGTUGA Moscow State Technical University of Civil Aviation) No. 199:11–22

Nesterenko GI (2010) Resurs i srok sluzhby konstruktsii samolyotov (Mechanical life and service life of airplane structures). Problemy mashinostroeniya i nadezhnosti mashin (Problems of mechanical engineering and machine reliability) No. 1

Nikolskaya P, Yakoreva A, Mironenko P, Myazina E (2015) Issledovanie RBK: na chem letaet Rossiya (RBC research: what Russia flies on). RBK. Obschestvo (RBC. Society), 7 Nov 2015. https://www.rbc.ru/research/society/27/11/2015/564de81a9a79472dab71463a. Accessed 27 Aug 2020

NP-001-15 Federalnye normy i pravila v oblasti ispolzovaniya atomnoi energii. Obschie polozheniya obespecheniya bezopasnosti atomnykh stantsii (Federal regulations and rules in the field of atomic energy use. General provisions for ensuring safety of nuclear power plants). Federal Environmental, Technological and Nuclear Supervision Service of Russia, Moscow, 2016

NP-003-97 (PNAE G-5-40-97) Federalnye normy i pravila v oblasti ispolzovaniya atomnoi energii. Trebovaniya k polnomasshtabnym trenazheram dlya podgotovki operatorov blochnogo punkta upravleniya atomnoi stantsii (Federal regulations and rules in the field of atomic energy use. Requirements for full-scale simulators for training operators of nuclear power plant unit control station). Scientific and Engineering Centre for Nuclear and Radiation Safety, Moscow, 2000

NP-034-15 Federalnye normy i pravila v oblasti ispolzovaniya atomnoi energii. Pravila fizicheskoi zaschity radioaktivnykh veschestv, radiatsionykh istochnikov i punktov khraneniya (Federal regulations and rules in the field of atomic energy use. Rules of physical protection of radioactive substances, radiation sources and storage facilities). Federal Environmental, Technological and Nuclear Supervision Service of Russia, Moscow, 2015

NP-073-11 Federalnye normy i pravila v oblasti ispolzovaniya atomnoi energii. Pravila fizicheskoi zaschity radioaktivnykh veschestv i radiatsionykh istochnikov pri ikh transpotirovanii (Federal regulations and rules in the field of atomic energy use. Rules of physical protection of radioactive substances and radiation sources during their transportation). Scientific and Engineering Centre for Nuclear and Radiation Safety, Moscow, 2012.

NP-084-15 Federalnye normy i pravila v oblasti ispolzovaniya atomnoi energii. Pravila kontrolya osnovnogo metalla, svarnykh soedinenii i naplavlennykh poverkhnostei pri ekspluatatsii oborudovaniya, truboprovodov i drugikh elementov atomnykh stantsii (Federal regulations and rules in the field of atomic energy use. Rules of controlling base metal, welded joints and claddings during the operation of equipment, pipelines and other elements of nuclear power plants). Federal Environmental, Technological and Nuclear Supervision Service of Russia, Moscow, 2016

Patrushev VS, Pletnyov IV, Spirochkin YK (1989) Aviatsionno-kosmicheskii trenazher (Aviation and space simulator). Patent SU 1,799,479, 24 Nov 1989

PNAE G-7-002-86 Normy raschyota na prochnost oborudovaniya i truboprovodov atomnykh energeticheskikh ustanovok (Strength analysis standards for the equipment and pipelines of nuclear power installations). Gosatomenergonadzor of the USSR. Energoatomizdat, Moscow, 1989

Podlinyaev OL, Karimov AA (2018) Psikhologicheskie osobennosti povedeniya lyudei v tolpe i ikh uchet sotrudnikami pravookhranitelnykh organov pri provedenii massovykh meropriyatii (Psychological features of the behavior of people in a crowd and their accounting by law enforcement officers during mass events). Psikhopedagogoka v pravookhranitelnykh organakh (Psychopedagogy in law enforcement agencies) 3(74):10–16

Predtechensky VM, Milinsky AI (1979) Proektirovanie zdanii c uchetom organizatsii dvizheniya lyudskikh potokov: Uchebn. posobie dlya vuzov. 2 izd., dop. i pererab. (Design of buildings taking into account organization of the movement of human flows: Textbook for universities, 2nd edn, supplemented and revised). Stroiizdat, Moscow

Rakitsky YV, Ustinov SM, Chernorutsky IG (1979) Chislennye metody resheniya zhyostkikh sistem (Numerical methods for solving stiff systems). Nauka, Moscow

RB-100-15 Rukovodstvo po bezopasnosti pri ispolzovanii atomnoi energii «Rekomendatsii po poryadku vypolneniya analiza nadezhnosti sistem i elementov atomnykh stantsii, vazhnykh dlya bezopasnosti, i ikh funktsii» (Safety guideline for the use of atomic energy. Recommendations on the procedure for performing reliability analysis of systems and elements of nuclear power plants important for safety, and their functions). Federal Environmental, Technological and Nuclear Supervision Service of Russia, Moscow, 2015

RB-130-17 Rukovodstvo po bezopasnosti pri ispolzovanii atomnoi energii «Polozhenie po ustanovleniyu urovnei fizicheskoi zaschity radiatsionnykh ob'ektov» (Safety guideline for the use of atomic energy. Provision on the establishment of levels of physical protection for radiation facilities). Federal Environmental, Technological and Nuclear Supervision Service of Russia, Moscow, 2017

RB-166-20 Rukovodstvo po bezopasnosti pri ispolzovanii atomnoi energii «Rekomendatsii po otsenke pogreshnostei i neopredelyonnostei rezultatov raschyotnykh analizov bezopasnosti atomnykh stantsii» (Safety guideline for the use of atomic energy. Recommendations for assessing errors and uncertainties in results of safety analyses of nuclear power plants). Federal Environmental, Technological and Nuclear Supervision Service of Russia, Moscow, 2020.

RIA Novosti (2020) Rogozin otsenil tekhnicheskoe sostoyanie MKS (Rogozin estimated the ISS technical condition). RIA Novosti, February 21, 2020. https://ria.ru/20200221/1565047329.html. Accessed 6 July 2020

RMRS (2022) Rossiiskii Morskoi Registr Sudokhodstva. Pravila po oborudovaniyu morskikh sudov. Chast II. Spasatelnye sredstva. ND № 2-020101-153-E (Russian Maritime Register of Shipping. Rules for the equipment of sea vessels. Part II. Life-saving appliances. Regulatory document No. 2-020101-153-E). RMRS, Saint Petersburg, 2022

Rosatom (2020) Sovremennye reaktory rossiiskogo dizaina. Reaktor WWER-1200 (Rosatom. Modern reactors of Russian design. Reactor WWER-1200). https://rosatom.ru/production/design/sovremennye-reaktory-rossiyskogo-dizayna/index.php?sphrase_id=1438146. Accessed 6 July 2020

SanPiN 2.6.1.24-03 (SP AS-03) Sanitarno-epidemiologicheskie Pravila i Normativy «Sanitarnye pravila proektirovaniya i ekspluatatsii atomnykh stantsii» (Sanitary and epidemiological rules and regulations. Sanitary rules for the design and operation of nuclear power plants). Ministry of Health of the Russian Federation, Moscow, 2003

SanPiN 2.6.1.2523-09 (NRB-99/2009) Sanitarno-epidemiologicheskie Pravila i Normativy «Normy radiatsionnoi bezopasnosti» (Sanitary and epidemiological rules and regulations. Radiation safety standards). Federal Center for Hygiene and Epidemiology of Rospotrebnadzor, Moscow, 2009

Shakhmatov AV (2012) Uchet sotsialno-psikhologicheskikh protsessov v mestakh massovogo skopleniya lyudei v operativno-sluzhebnoi deyatelnosti sotrudnikov OVD (Accounting of socio-psychological processes in crowded places in the operational and service activities of police officers). Vestnik Sankt-Peterburgskogo universiteta MVD Rossii (Bulletin of the Saint Petersburg University of the Ministry of Internal Affairs of Russia) 1(53):268–273

Sidorkova I, Lindell D, Bogachyov A (2018) «Poskosmos» vs SpaceX: kto pobezhdaet v kosmicheskoi gonke (Roscosmos vs SpaceX: who wins the space race). RBK. Tekhnologii i media (RBC. Technologies and media), 8 Feb 2018. https://www.rbc.ru/technology_and_media/08/02/2018/5a7b1b5a9a7947a1973ea3b8. Accessed 31 Marh 2022

Sorokin LA (2017) Informatsionno-analiticheskaya podderzhka upravleniya bezopasnost'yu v mestakh massovogo prebyvaniya lyudei (Information and analytical support of safety management in places of mass presence of people). Dissertation. Academy of the State Fire Service of the Ministry of Emergency Situations of Russia, Moscow

Sozinov A (2022) SpaceX zapustit 60 raket Falcon 9 v etom godu—eto bolshe, chem Kitai zapustil za 2021 god (SpaceX will launch 60 Falcon 9 rockets this year—that is more than China launched in 2021). 3D News, 29 March 2022. https://3dnews.ru/1063012/spacex-zapustit-60-raket-falcon-9-v-etom-godu-eto-bolshe-chem-kitay-zapustil-za-2021-god. Accessed 31 March 2022

SP 1.13130.2009 Svod Pravil. Sistemy protivopozharnoi zaschity. Evakuatsionnye puti i vykhody (s Izmeneniem № 1, utverzhdennym i vvedennym v deistvie s 01.02.2011 Prikazom MChS Rossii ot 09.12.2010 № 639) (Code of rules. Systems of fire protection. Evacuation ways and exits with Change No. 1 approved by the Directive of the Ministry of Emergency Situations of Russia dated 09.12.2010 No. 639 and effective since 01.02.2011). Ministry of Emergency Situations of Russia, Moscow, 2009

SP 3.13130.2009 Svod Pravil. Sistemy protivopozharnoi zaschity. Sistema opoveshcheniya i upravleniya evakyatsiei lyudei pri pozhare. Trebovaniya pozharnoi bezopasnosti (Code of rules. Systems of fire protection. System of annunciation and management of human evacuation at fire. Requirements of fire safety). Ministry of Emergency Situations of Russia, Moscow, 2009

SP 117.13330.2011 (SNiP 31-05-2003) Svod Pravil. Stroitelnye Normy i Pravila Rossiiskoi Federatsii. Obschestvennye zdaniya administrativnogo naznacheniya (Code of rules. Building regulations and codes of the Russian Federation. Public administrative buildings). Standartinform, Moscow, 2006

SP 151.13330.2012 Svod Pravil. Inzhenernye izyskaniya dlya razmescheniya, proektirovaniya i stroitelstva AES. Chasti I i II (Code of rules. Engineering surveys for placement, design and construction of nuclear power plants. Parts I and II). Federal Agency for Construction and Housing and Communal Services, Moscow, 2012

SP 165.1325800.2014 Svod Pravil. Inzhenerno-tekhnicheskie meropriyatiya po grazhdanskoi oborone. Aktualizirovannaya redaktsiya SNiP 2.01.51-90 (s Izmeneniem № 1, utverzhdennym i vvedennym v deistvie Prikazom Minstroya Rossii ot 24 oktyabrya 2017 goda № 1471/pr s 25.04.2018, i Izmeneniem № 2, utverzhdennym i vvedennym v deistvie Prikazom Minstroya Rossii ot 26 noyabrya 2020 goda № 725/pr s 27.05.2021) (Code of rules. Engineering and technical measures for civil defense. Updated version of SNiP 2.01.51-90 with Change No. 1 approved by the Directive of Minstroi of Russia dated October 24, 2017 No. 1471/pr and effective since 25.04.2018, and Change No. 2 approved by the Directive of Minstroi of Russia dated November 26, 2020 No. 725/pr and effective since 27.05.2021). Minstroi of Russia, Moscow, 2014

SP 255.1325800.2016 Svod Pravil. Zdaniya i sooruzheniya. Pravila ekspluatatsii. Osnovnye polozheniya (Code of rules. Buildings and structures. Operating rules. General provisions). Minstroi of Russia, Moscow, 2016

SP 309.1235800.2017 Svod Pravil. Zdaniya teatralno-zrelischnye. Pravila proektirovaniya (Code of rules. Theater and entertainment buildings. Design rules). Minstroi of Russia, Moscow, 2016

Spirochkin YK (1987) Metodika raschyota dinamicheskogo nagruzheniya gruzhov pri avariinoi posadke letatelnogo apparata (Method for analyzing dynamic loads on items of mass during an aircraft emergency landing). Dissertation. NPO *Energia*, Kaliningrad, Moscow region, 1987 (Defended in the Central Institute of Mechanical Engineering (TsNIImash), Kaliningrad, Moscow region, 1988)

Spirochkin YK (1988) Iskhodnye dannye na provedenie eksperimentalnykh issledovanii parametrov reaktsii posadochnoi poverkhnosti s ispolzovaniem zhyostkikh maketov. Otchet P 26478-017 (Baseline data for experimental studies on the landing surface reaction parameters using rigid mockups. Report P 26478-017). NPO *Energia*, Kaliningrad, Moscow region

Spirochkin YK (1993) Komp'yuternoe modelirovanie dinamiki konstruktsii pri avariyakh (Computer simulation of structural dynamics in accidents). Matematicheskoe modelirovanie (Math Models Comput Simul) 5(6):85–103

Spirochkin YK (1994) Spetsialnye konechnye elementy dlya zadach dinamiki konstruktsii (Special finite elements for structural dynamics problems). Matematicheskoe modelirovanie (Math Models Comput Simul) 6(8):85–91

Spirochkin YK (1998) Konechnoelementnoe modelirovanie dinamiki obolochek, vzaimodeistvuyuschikh s tonkimi sloyami zhidkosti (Finite element modeling and simulation of the dynamics of shells interacting with thin layers of fluid). Prikladnye problemy prochnosti i plastichnosti. Chislennoe modelirovanie fiziko-mekhanicheskikh protsessov: Mezhvuzovskii sbornik (Applied problems of strength and plasticity. Numerical modelling of physical and mechanical processes: Interuniversity collection), 58:110–121

Spirochkin YK (2019) Bezopasnost rossiiskikh AES s tochki zreniya inzhenera-mekhanika (Safety of Russian nuclear power plants from the viewpoint of mechanical engineer). SUPER Publishing House, Saint Petersburg

Spirochkin YK (2020) Utrata proektnykh znanii (Loss of design knowledge). Aviapanorama—Mezhdunarodnyi aviatsionno-kosmicheskii zhurnal (Aviapanorama—International Aerospace Magazine), 22 July 2020. https://www.aviapanorama.ru/2020/07/utrata-proektnyh-znanij. Accessed 22 July 2020

Spirochkin YK, Saprykin OA (2021) Kosmicheskii polyot pod solnechnym parusom (Space flight under a solar sail). AviaSafety.ru, 6 Jan 2021. https://aviasafety.ru/36409. Accessed 7 Jan 2021

Spirochkin YK, Shenk YV (1990) Chislennoe modelirovanie avariinoi posadki vertoleta (Numerical simulation of a helicopter emergency landing). Problemy udara, razrusheniya i tekhnologii (Problems of impact, destruction and technology) 10(4):62–63 (Publishing house VNITsUR-Resurs, Riga)

STO 1.1.1.04.001.1500-2018 Standart organizatsii. Pravila pozharnoi bezopasnosti pri ekspluatatsii atomnykh stantsii (Standard of organization. Fire safety rules for the operation of nuclear power plants). JSC Concern Rosenergoatom, Moscow, 2018

The Universe (2020) Puskovaya statistika 2019 g. (Launch statistics 2019), 1 Jan 2020. https://universemagazine.com/ru/puskovaya-statystyka-2019-g. Accessed 31 March 2022

Timofeev AG, Zlobin PV (2015) Kontseptsiya «bezlyudnogo» proizvodstva (The concept of "unmanned" production). Izvestiya Rossiiskogo ekonomicheskogo universiteta im. G. V. Plekhanova (Proc G.V. Plekhanov Russ Univ Econ) 4(22):388–399

Timoshenko DM (2014) Metody avtomaticheskoi identifikatsii lichnosti po izobrazheniyam lits, poluchennym v nekontroliruemykh usloviyakh (Methods of automatic identification of a person based on images of faces obtained in uncontrolled conditions). Dissertation. Petrozavodsk State University, Petrozavodsk

TR TS 001/2011 Tekhnicheskii reglament Tamozhennogo Soyuza «O bezopasnosti zheleznodorozhnogo podvizhnogo sostava» (Technical Regulations of the Custom Union: On the safety of railway rolling stock). Moscow, 2011

TR TS 002/2011 Tekhnicheskii reglament Tamozhennogo Soyuza «O bezopasnosti vysokoskorostnogo zheleznodorozhnogo transporta» (Technical Regulations of the Custom Union: On the safety of high-speed rail transport. Moscow, 2011

TR TS 018/2011 Tekhnicheskii reglament Tamozhennogo Soyuza «O bezopasnosti kolesnykh transportnykh sredstv» (Technical Regulations of the Customs Union: On the safety of wheeled vehicles. Moscow, 2011 as amended on 21 June 2019)

Zhuravlev AL, Sosnin VA (2014) Psikhologiya massovogo povedeniya: istoki i sovremennye tendentsii issledovaniya (Psychology of mass behavior: the origins and contemporary research tendencies). Znanie. Ponimanie. Umenie (Knowl Underst Skills) 1:49–61